04-HHE-7

An Investigation into Biosolids Sampling and Handling Methods for U.S. EPA-Approved Microbial Detection Techniques

by:

Sharon C. Long
University of Wisconsin

Victor D'Amato
Tetra Tech, Inc.

Anita Bahe
Lynx Group International, LLC

2009

WERF

The Water Environment Research Foundation, a not-for-profit organization, funds and manages water quality research for its subscribers through a diverse public-private partnership between municipal utilities, corporations, academia, industry, and the federal government. WERF subscribers include municipal and regional water and wastewater utilities, industrial corporations, environmental engineering firms, and others that share a commitment to cost-effective water quality solutions. WERF is dedicated to advancing science and technology addressing water quality issues as they impact water resources, the atmosphere, the lands, and quality of life.

For more information, contact:
Water Environment Research Foundation
635 Slaters Lane, Suite 300
Alexandria, VA 22314-1177
Tel: (703) 684-2470
Fax: (703) 299-0742
www.werf.org werf@werf.org

This report was co-published by the following organizations. For non-subscriber sales information, contact:

IWA Publishing
Alliance House, 12 Caxton Street
London SW1H 0QS, United Kingdom
Tel: +44 (0) 20 7654 5500
Fax: +44 (0) 20 7654 5555
www.iwapublishing.com
publications@iwap.co.uk

Library of Congress Catalog Card Number: 2007941991
IWAP ISBN: 978-1-84339-520-1/1-84339-520-7

Wisconsin State Laboratory of Hygiene-University of Wisconsin, ARCADIS U.S., Inc., Tetra Tech, Inc. University of Massachusetts

The research on which this report is based was developed, in part, by the United States Environmental Protection Agency (EPA) through Cooperative Agreement No. CR-83155901 with the Water Environment Research Foundation (WERF). However, the views expressed in this document are solely those of Wisconsin State Laboratory of Hygiene-University of Wisconsin, ARCADIS U.S., Inc., Tetra Tech, Inc. University of Massachusetts and neither EPA nor WERF endorses any products or commercial services mentioned in this publication. This report is a publication of WERF, not EPA. Funds awarded under the Cooperative Agreement cited above were not used for editorial services, reproduction, printing, or distribution.

ACKNOWLEDGMENTS

Report Preparation

Principal Investigator:

Sharon C. Long, Ph.D.
Wisconsin State Laboratory of Hygiene and University of Wisconsin
(previously with the University of Massachusetts)

Co-Principal Investigator:

Victor D'Amato, P.E., MSEE
Tetra Tech, Inc. (previously with ARCADIS U.S., Inc.)

Project Team:

Anita Bahe, Ph.D.
Glenn Cassidy, Ph.D.
Lynx Group International, LLC

Leah D. Cronan
Nicholas S. Keuler
Kristine L. Lund
Jun Zhang
University of Wisconsin

Justin DeMello
Cora Olson
University of Massachusetts

Robert Rubin, Ph.D., Professor Emeritus
North Carolina State University

James P. Scisson, Jr.
URS Consultants (previously with ARCADIS U.S., Inc.)

Project Subcommittee

William Cairns, Chair, Ph.D.
Trojan Technologies, Inc.

Randi McCuin
Clancy Environmental Consultants, Inc.

Nancy Penney
Water Corporation of Western Australia

Geeta Rijal, Ph.D.
Metropolitan Water Reclamation District of Greater Chicago

Daniel R. Shelton, Ph.D.
USDA/ARS/NRI/ECL

James Smith, Jr., D.Sc.
U.S. EPA-NMRL

Prakasam Tata, Ph.D., Q.E.P.
Tata Associates International

Water Environment Research Foundation Staff

Director of Research:	Daniel M. Woltering, Ph.D.
Program Manager:	Alan Hais, P.E., BCEE
	Lola Olabode, M.P.H.

ABSTRACT AND BENEFITS

Abstract:

The objective of this research was to develop scientifically defensible methods of collecting and handling representative biosolids samples that are used for microbial analysis. Investigators conducted three phases of research.

In the first phase, researchers established the status of sampling guidance and practices for biosolids. They also defined the research questions targeted in subsequent phases of the research.

In the second phase, the researchers sampled and analyzed sludge and biosolids products from four facilities utilizing different solids handling and treatment technologies. The purpose was to determine how various sample collection and handling parameters influence sample integrity and representativeness, based on research questions identified in Phase I.

In the third phase, the team conducted field testing at nine utilities to confirm the performance of the sampling and handling strategies. They also compared microbial monitoring results between this project and the methodologies of the host utilities, whose protocols were examined to determine the suitability of their sampling approach(es).

This research demonstrated that analysis of multiple, discrete, grab samples provides insight into product variability. Proper handling and adherence to sample size and storage protocols consistent with the Part 503 Regulations provides reliability when measuring biosolids microbial content from the biosolids production process being sampled. The project demonstrated that sampling and analysis of *each* biosolids batch would boost utility and stakeholder confidence in the microbial safety of the product. For most utilities, this would be more frequently than the 503 Regulation mandated sampling and testing schedules that are based on dry ton biosolids production.

Benefits:

- provides recommendations for collecting representative samples of biosolids for testing, using EPA-approved microbial analysis techniques;
- provides a foundation for sampling and testing practices that minimize potential health risks by providing the means to identify when non-Class A or Class B-compliant pockets of biosolids are present; and
- presents an interpretation of Part 503 biosolids sampling guidance and practices to allow utilities to make informed choices when designing a sampling strategy that fits their facility's processes and material flow configuration.

Keywords: Biosolids, sludge, handling, microbial testing, sampling.

TABLE OF CONTENTS

LIST OF TABLES

LIST OF FIGURES

LIST OF ACRONYMS

CFU	Colony forming unit
DAFT	Dissolved air floatation tank
EMS	Environmental management system
GBT	Gravity belt thickener
LIA	Lysine iron agar
LTB	Lauryl tryptose broth
MGD	Million gallons per day
MPN	Most probable number
MSE	Mean square error
PFRP	Process to further reduce pathogens
PFU	Plaque forming unit
PI	Principal investigator
POTW	Publicly owned treatment works
PSC	Project subcommittee
PSRP	Process to significantly reduce pathogens
QA/QC	Quality assurance/quality control
RAS	Recycled activated sludge
SOUR	Specific oxygen uptake rate
SRT	Solids residence time
TPAD	Temperature phased anaerobic digestion
TSI	Triple sugar iron agar
U.S. EPA	U.S. Environmental Protection Agency
VAR	Vector attraction reduction
WAS	Waste activated sludge

Executive Summary

Biosolids are highly processed materials produced from the solid residual by-products (sludge) of wastewater treatment operations and often can be used as a fertilizer or soil amendment. Depending on a number of variables, including the sources of wastewater in the service area of the public utility or treatment works, untreated sludge may contain various levels of undesirable compounds, such as toxic metals and/or pathogens. Such biosolids, if land applied, can potentially result in an environmental and/or public health risk.

Protecting human health and the environment from these potential risks depends in part on monitoring to detect and quantify pollutants in sewage sludge and processed/treated solids (i.e. biosolids). Although the precision and accuracy of laboratory methods used to quantify constituents in biosolids has been a subject of emphasis, errors and variability in the data that result from differences in sampling and handling practices can often be overlooked. Currently, available guidance is somewhat inconsistent and often contains conflicting details or offers little detail as to the specifics of proper sample collection and handling procedures for measuring microbial content in biosolids materials.

The objective of this research was to develop scientifically defensible methods of collecting and handling representative biosolids samples that are used for microbial analysis, consistent with the Part 503 Regulations. As such, the project was designed to: a) examine the current practices for sampling and handling biosolids or processed sludge and determine the effectiveness of regulatory compliance via microbial testing; b) research strategies for obtaining representative samples; and c) make recommendations and prepare guidance concerning the best approaches for sampling and handling three types of biosolids materials – liquid, cake and compost – to best assure protection of human health and the environment, and to promote regulatory and stakeholder confidence.

To accomplish these objectives the project was conducted in three phases:

- Phase I involved a literature review, a survey of a representative population of wastewater treatment operations, and an experts' workshop.
- Phase II engaged four facilities, each providing sludge or biosolids produced using different handling and/or treatment technologies that were subject to laboratory experiments which addressed sampling and handling questions derived during Phase I.
- Phase III was the field testing of sampling and handling approaches identified in Phase II. The researchers compared results and practices employed at eleven participating facilities.

Finally, a series of communications documents was prepared and beta tested by a number of wastewater utility operators and managers. These communications tools were designed to convey the importance of sampling and handling details at multiple stakeholder levels.

This project generally revealed that sample collection practices vary widely among wastewater utilities, owing to a host of influences including: inconsistent and/or unclear guidance specific to biosolids sampling; variable sources of guidance and regulatory requirements for sampling (state, federal, laboratory, utility specific permits, etc); level of staff training; effectiveness of communications; and facility management (budgets, internal programs, treatment processes and biosolids end use).

This project also revealed insights into the practicality and applicability of various sampling and handling protocols. Examples include: the necessity for purging pipes/spigots prior to collecting liquid samples and associated alternatives; frequency and location of sample collection relative to data quality; effectiveness of sampling tools and containers for both collection and handling prior to analysis; implications of proper sample containment, storing, and cooling procedures; sterilization procedures; implications of sample composites versus discrete samples in terms of representativeness of samples for biosolids that have been stored; and suitability of multiple microbial measures.

In brief, multiple discrete samples collected from *each* batch of biosolids should be collected and analyzed for microbial content. The team recommends three to ten samples, if available staff and budgets allow. One liter samples should be collected in sterile containers and placed on ice immediately by having a cooler available at the sampling point. For process monitoring, samples should be collected after the last treatment process, when the biosolids are being mixed or are in motion, i.e., as liquid or cake is transported to storage. Sample analysis should begin as close in time to sample collection as possible.

The importance of highly effective, accurate communication was a point of emphasis during the survey and beta-test activities of this project. In developing communication products, the team was reminded of the challenges and opportunities in conveying technical, regulatory and science-based information to extremely diverse audiences, particularly where high performance and public confidence are goals.

Disclaimer

The recommendations from this work are meant to be used in conjunction with EPA, state, and local regulations, but not to substitute for those regulations. Any apparent inconsistencies are unintended and the appropriate regulations take precedent.

CHAPTER 1.0

INTRODUCTION AND OBJECTIVES

1.1 Introduction

Advancing the principles of sustainability takes on many forms, including the beneficial use of materials sometimes viewed as wastes of little value – like sewage sludge, a by-product of wastewater treatment processes. Biosolids are highly processed and treated sewage sludge intended for beneficial use. Biosolids materials can have value as soil conditioners, providing organic matter to improve soil characteristics. They can also be an organic nutrient source for agriculture and garden purposes. Despite this value, some consumers and other stakeholders have concerns about the presence of hazardous constituents (i.e. pathogens, metals and other toxics) in biosolids products. Accordingly, the use of these products is highly regulated. Despite this, biosolids use is sometimes met with public resistance.

The Part 503 Regulations contain criteria used to classify biosolids products as Class A or Class B with respect to microbial or pathogen content. Class A products have limited potential for pathogen exposure as they require treatment by a "process to further reduce pathogens" (PFRP) and typically contain very low levels of measured pathogens and indicator organisms. Class B products are typically treated by a "process to significantly reduce pathogens" (PSRP) and, as such, their use on agricultural lands may require site and harvesting restrictions to assure protection of public health.

In order to determine whether a biosolids product meets Class A or Class B definitions with respect to microbial (pathogen) content, the treatment process must be tightly monitored and the products sampled and tested. This constitutes a "two-barrier" approach to assuring the microbial safety of biosolids products. Subpart D of Part 503 of the Code of Federal Regulations, commonly known as the "503 regulations," specifies various methods that can be used to monitor for enteric viruses, fecal coliform, *Salmonella* sp. bacteria, viable helminth ova, specific oxygen uptake rate (SOUR), solids (total, fixed, and volatile), and percent volatile solids reduction (40 CFR 503, 1993).

Because sampling and analysis practices are an important part of the “two-barrier” approach, this research assessed the current effectiveness of methods and guidelines and informs potential modifications to sampling and handling practices. This project utilized a combination of existing information synthesis, experimental work, and assessment of the state of practice to develop recommendations for sampling and handling of samples intended for microbial analysis. The researchers considered technical, scientific, economic and practical application implications.

1.2 Objectives and Research Approach

The objective of this research was to develop scientifically defensible methods of collecting and handling representative biosolids samples that are used for microbial analysis, consistent with the Part 503 Regulations. When gathered in a consistent and representative manner, the data aim to provide information that protects human health and the environment from potential release of pathogens from land-applied biosolids.

To accomplish these objectives the project was conducted in three phases. The first phase gathered information and established the status of sampling guidance and practices for biosolids. In Phase II, researchers sampled and analyzed treated sludge or biosolids products from four target facilities utilizing different solids handling and treatment technologies. They determined which sample collection and handling parameters most affect sample integrity and representativeness. In Phase III, the team conducted field testing of the sampling and handling protocols tentatively developed in earlier phases. Finally, a series of communications documents were prepared and beta tested by a number of wastewater utility operators and managers. These communications tools were designed to convey the importance of sampling and handling details at multiple stakeholder levels.

CHAPTER 2.0

STATE OF BIOSOLIDS SAMPLING PRACTICE

In Phase I of the project, three major sources of information established the basis for existing biosolids sampling and sample handling practices:

- a literature search on standard sampling practices across disciplines including wastewater, biosolids, food, soils, solid waste and hazardous waste;
- a two-tiered (manager and staff) survey of biosolids sampling practices in place at wastewater utilities; and
- a workshop of biosolids and testing experts conducted as part of WEFTEC05.

This base of information was also used to formulate the research questions that were addressed in Phase II and III of this project. A synopsis of the state of understanding and practice with respect to sampling is presented below.

2.1 Literature Review

2.1.1 Background

Biosolids are the highly treated and processed solids derived from domestic wastewater treatment processes. Biosolids are a recycled resource that has many potential environmental, economic and agricultural benefits. The beneficial use of biosolids typically is as a soil amendment and can range from small bags for personal garden use to large-scale agricultural applications (U.S. EPA, 1994). Some of the benefits of adding biosolids to soils are that they supply nutrients (N and P, among others), replenish soil organic matter, and increase the soil's water-holding capacity (U.S. EPA, 2000a). The use of biosolids is limited by public perception that they are a human waste-derived product and therefore are inherently unsafe. However, biosolids are a highly processed organic product produced by wastewater treatment facilities that can be beneficially recycled. To assure minimal risk to human health and the environment, the EPA mandates adherence to stringent regulations and guidance primarily contained within the "Part 503" or "the Part 503 rule" (40 CFR 503, 1993) when producing biosolids intended for land application.

Land application of highly treated, quality assured biosolids is a greener, more-sustainable approach to wastewater residuals management than disposal (U.S. EPA, 2000a), which usually involves incineration or landfilling (U.S. EPA, 1994). This study focused on biosolids products intended for land application, particularly liquid, cake and composted biosolids materials.

Liquid and cake biosolids may be produced by a variety of aerobic or anaerobic stabilization processes. Liquid biosolids are typically comprised of up to approximately 7% total solids (U.S. EPA, 2000b). The volume of product can be reduced by extracting the excess water from the material through the processes of draining, centrifuging or pressing (U.S. EPA, 2000a). The resulting dewatered products – cake material – may be up to 25% solids. Composting is an aerobic stabilization process in which bulking agents – typically organic matter such as wood chips or yard waste – are added to the wastewater residuals. The total solids contents of composted biosolids can vary, but may be as high as 40 to 60% depending on the bulking agents used (U.S. EPA, 2000a).

The biosolids starting materials – sewage sludge – may contain potentially harmful constituents such as heavy metals, pathogens or toxic organics depending on the characteristics of the treatment plant's service area. These constituents may pose risks to exposed populations (U.S. EPA, 1995). Strict treatment process monitoring and periodic testing of biosolids products, as required under the 503 Regulations, are intended to determine that potentially harmful constituents are at or below acceptable levels in the final biosolids product. The constituents of particular interest to this project are microorganisms, specifically pathogens. Microorganisms of concern in biosolids include bacterial, viral, and parasitic human pathogens that can be shed by infected individuals in feces and urine that subsequently survive through the treatment processes and end up in the solids. Biosolids must meet strict requirements before they are deemed suitable for land application. As indicated above, these strict requirements have been put into place to minimize associated risks and ensure that land application of processed wastewater residuals remains a sustainable and socially acceptable recycling practice.

There are two classes of biosolids based on pathogen removal, Class A and Class B (U.S. EPA, 1994). The difference between the two classes is the type of treatment that the biosolids material is subject to prior to application - processes to further reduce pathogens (PFRP) versus processes to significantly reduce pathogens (PSRP), for Class A and Class B products, respectively. The processes that meet those definitions are described in the 503 Regulations. To be considered Class A or B biosolids, the material must meet the ceiling concentrations for microbial indicators and pathogens as indicated in Table 2-1.

Table 2-1. Part 503 Microorganism Density Limits for Biosolids Products.

Indicator or Pathogen	Density Limits (dry wt basis)
Class A	
Salmonella	<3 MPN/4 g or
Fecal coliforms	<1000 MPN/g and
Enteric viruses	<1 PFU/4 g and
Viable helminth ova	<1 ova/4 g
Class B	
Fecal coliforms	<2,000,000 MPN/g

MPN - most probable number

Class A biosolids products must be treated using at least one of six alternatives for achieving PFRP as well as be tested to ensure that fecal coliform or *Salmonella* density requirements are met. Pathogen reduction alternatives are listed in Table 2-2. There are few federal government restrictions on the use of Class A biosolids, thus they can be used for many applications with assumed minimal risk to public health and the environment.

For Class B products, pathogen reduction requirements are demonstrated by meeting one of three alternatives as presented in Table 2-2. The only microbial density monitoring limit for Class B biosolids is that they contain less than 2 million fecal coliforms per gram dry weight

(Table 2-1). Class B biosolids potentially contain only very low levels of pathogens but are restricted to non-public, non-harvesting or grazing land uses to reduce the likelihood of human or animal exposure to pathogens (U.S. EPA, 1994).

Table 2-2. Part 503 Class A and Class B Pathogen Reduction Requirements.

To qualify as **Class A**, the biosolids must be treated using at least one of the following six alternatives as well as meet the fecal coliform or *Salmonella* specific density requirements. *Alternative 1 – Thermally Treated Biosolids* ▪ Use one of four time-temperature regimens *Alternative 2 – Biosolids Treated in a High pH-High Temperature Processes* ▪ Specifies pH, temperature, and air-drying requirements *Alternative 3 – For Biosolids Treated in Other Processes* ▪ Demonstrates that the process can reduce enteric viruses and viable helminth ova. Maintain operating conditions used in the demonstration. *Alternative 4 – Biosolids Treated in Unknown Processes* ▪ Test for pathogens at each time the biosolids are used or disposed of or are prepared for sale or to be given away. *Alternative 5 – Use of PFRP* ▪ Biosolids are treated by one of the Processes to Further Reduce Pathogens (PFRP) *Alternative 6 – Use of a Process Equivalent to PFRP* ▪ Biosolids are treated in a process equivalent to one of the PFRPs, as determined by the permitting authority.
To qualify as **Class B**, the biosolids must be treated using at least one of the following three alternatives. *Alternative 1 – Monitoring of Indicator Organisms* ▪ Test for fecal coliform density as an indicator for all pathogens at the time of biosolids use or disposal *Alternative 2 – Use of PSRP* ▪ Biosolids are treated by one of the Processes to Significantly Reduce Pathogens (PSRP) *Alternative 3 – Use of Process Equivalent to PSRP* ▪ Biosolids are treated in a process equivalent to one of the PSRPs, as determined by the permitting authority.

After Table 2-5 Summary of Class A and Class B Pathogen Reduction Requirements, U.S. EPA, 1994

2.1.2 Microbial Testing of Biosolids

A key component in maintaining biosolids products classifications is testing. Of particular focus for this project is microbial testing, using EPA-approved methods. Among the many issues regarding the ability of these tests to demonstrate biosolids product safety for land application are the microbial analyses themselves, as well as sampling techniques and sample handling.

The Federal Register and current Part 503 regulations include guidance on specific microbial analyses that can be conducted on biosolids samples to assure Class A or B microbial levels (U.S. EPA, 1994; 72 *Fed. Reg.* 57 (26 March 2007)). Table 2-1 summarizes the limits on microorganism densities for testing of biosolids products.

In order to determine whether a biosolids product meets Class A or Class B definitions, the process must be monitored and sampled. The biosolids production process must meet operational requirements, and the microbial testing must yield indicator and/or pathogen densities below the limits presented in Table 2-1. Part 503 Subpart D and the Federal Register specify the methods that can be used to monitor for enteric viruses, fecal coliform(s), *Salmonella* sp. bacteria, viable helminth ova, specific oxygen uptake rate (SOUR), solids (total, fixed, and volatile), and percent volatile solids reduction.

The allowed analytical methods are summarized in Table 2-3. The U.S. EPA recommends Methods 1680 and 1681 for measuring fecal coliforms in sewage sludge/biosolids, and Method 1682 for measuring *Salmonella* (72 *Fed. Reg.* 57 (26 March 2007)). Understanding the characteristics of applicable microbial testing methods can provide insights into the accuracy and precision required in sample collection and handling which in turn can help in designing guidance.

Table 2-3. Alternate Analytical Methods for Sewage Sludge Required Under Part 503.

Analysis	Methodology
Fecal Coliform	Standard Methods Part 9221E or Part 9222D (APHA et al., 2005) Method 1680 (recommended; U.S. EPA, 2006) Method 1681 (recommended; U.S. EPA, 2006a)
Enteric Viruses	ASTM Method D 4994-89 (ASTM, 1994)
Salmonella spp.	Standard Methods Part 9260D (APHA et al., 2005) Kenner and Clark (1974) Method 1682 (recommended; U.S. EPA, 2006b)
Viable Helminth Ova	Yanko (1987)
SOUR	Standard Methods Part 2710B (APHA et al., 2005)
Total, Fixed, and Volatile Solids	Standard Methods Part 2540G (APHA et al., 2005)
Percent Volatile Solids Reduction	Appendix H of EPA document "Control of Pathogens in Sewage Sludge" (U.S. EPA, 1999)

2.1.2.1 Fecal Coliforms

Fecal coliforms are a class of bacteria that are defined by the methods of analysis; they are capable of growth on defined media at 44.5°C in 24 hours (APHA et al., 2005). They can be found in water, soil or the digestive system of warm-blooded animals (U.S. EPA, 2003). With the exception of *E. coli* O157:H7, most fecal coliforms are typically not harmful to humans but they are used as an indicator of the possible presence of other bacteria with the potential to cause diseases such as cholera, dysentery and typhoid (U.S. EPA, 2003). Fecal coliforms can pose a risk of opportunistic infection to humans if they come in contact with the face, ears or open wounds (U.S. EPA, 2003). Fecal coliforms are used in testing for the microbial safety of biosolids because they are present in high numbers in raw wastewater. A reduction in the numbers of fecal coliforms may therefore correlate to a reduction in pathogens. Since fecal coliforms are ubiquitous in wastewater treatment facilities, any equipment or testing devices used to sample biosolids for fecal coliform testing need to be sterilized and may never be reused until they have been cleaned and disinfected.

One U.S. EPA-approved test method for fecal coliform is Standard Method 9222D using the membrane filter (MF) technique (APHA et al., 2005). The MF technique utilizes an enriched lactose media, m-FC, with incubation at 44.5°C for 24 ± 2 hours. Blue colonies are counted as fecal coliforms, while yellow or gray colonies are considered background. The culture dishes need to be sealed tight in waterproof bags in order to be submerged in a waterbath for incubation. The goal with respect to sample size is to select a sample volume for filtration that will result in less than 200 and more than 20 total colonies for each filter membrane. The optimal colony density should be between 20 and 60 per membrane. Challenges with enumerating solid materials, such as biosolids, include producing a suspension dilute enough so as not to clog or crowd the membrane with particulate material. Because of this, while U.S. EPA-approved and recommended for water samples, this procedure is not the preferred method for biosolids samples.

Another U.S. EPA-approved fecal coliform enumeration technique is Standard Method 9221E (APHA et al., 2005). This is a fermentation tube method that uses a most-probable-number (MPN) format and an EC medium. A series of five replicate tubes, each including at least three different inoculation volumes of presumptive media (lauryl tryptose broth, LTB), is used. A small inoculum from each positive tube (as indicated by gas production and growth) is transferred into an EC tube using a sterile wooden stick or transfer loop. The EC tubes are incubated for 24 hours at 44.5°C in a waterbath. A positive test is indicated by gas and growth (visible turbidity) in the tubes. The most probable number is calculated using Table 9221.IV in Standard Methods. Similar to the MF technique, the challenges in testing solid materials such as biosolids are in obtaining representative suspensions and dilutions for inoculation of the tubes, and choosing a series of test volumes that result in a combination of positive and negative tubes.

The U.S. EPA preferred fecal coliform methods are U.S. EPA Method 1680 or 1681. Method 1680 is a modification of Standard Method 9221E involving presumptive incubation in LTB at 36°C for 24 ± 2 hours or 48 ± 3 hours (U.S. EPA, 2006). For this method, at least four sample levels are required and five or more are preferred. At a minimum, replicates containing the equivalent of 1.0, 10^{-1}, 10^{-2}, and 10^{-3} g of original sample should be inoculated. To prepare sample dilutions, 30.0 g of solid material are weighed and placed in a blender with 270 mL of sterile dilution water. Through this dilution, one mL of homogenized sample will contain 10^{-1} g of original sample and serial dilutions are used to generate the other, more dilute replicates. However, dilution errors may be compounded, and thus potentially compromise the integrity and representativeness of dilute replicates.

The second EPA preferred fecal coliform enumeration method, Method 1681, is an adaptation of Standard Method 9221E using A-1 medium in a direct, single-step test (U.S. EPA, 2006a). As with Method 1680, this test is performed in a most-probable-number format. Four different sample volume inoculum levels are required, and five or more are preferred. Each sample volume is inoculated into five replicate tubes and incubated at 35°C for 3 hours, then transferred to 44.5°C for 21 ± 2 hours. Tubes that exhibit growth (turbidity) and gas are considered positive for fecal coliforms. As above, a solid sample is homogenized using sterile dilution water. One challenge in testing solids is producing representative replicates at each sample volume level.

Both proposed EPA methods, 1680 and 1681 performed in MPN-mode, result in bacterial density estimates for which the 95% confidence limits span an order of magnitude. Thus, ideally, sampling plans should be designed in a way that captures a variance of significantly less than an order of magnitude. In practice, however, this would be difficult to ascertain, since the individual components of the error estimate are difficult, at best, to characterize.

2.1.2.2 *Salmonella*

Salmonella are a genus of bacteria that account for the majority of human gastrointestinal disease of known etiology in the United States (U.S. FDA, 1992). The Food and Drug Administration reports that approximately two to four million cases of Salmonellosis occur annually in the United States. The pathogen is usually spread through ingestion of food, raw meats, water, insects and feces into the human body where the pathogen then colonizes the digestive tract.

When testing for *Salmonella* using Standard Method 9260D, large sample sizes are essential to detect small concentrations of the bacteria in biosolids (Kenner and Clark, 1974). The first step is to inoculate selective enrichment broth cultures, which can be completed using a sodium acid selenite, selenite cystine (24-hour 35°C), selenite-F (24-hour 35°C), or tetrathionate medium. Following appropriate incubation times for enrichment, the broth cultures are streaked onto plates of brilliant green and xylose lysine desoxycholate (XLD) agars and are incubated for 24 hours at 35°C. At least one, and preferably two or three colonies suspected of being *Salmonella* (pink-white colonies with red medium on brilliant green agar; red with black centers on XLD agar) are selected and transferred to triple sugar iron (TSI) and lysine iron agar (LIA) slants. These are incubated for 24 hours at 35°C. *Salmonella* spp. will produce a H_2S black precipitate and gas in TSI, and a purple butt with a H_2S black precipitate on LIA. This process can take up to five days for completion. Method 9260D can be conducted in MPN-mode by inoculating various volumes into various numbers of replicate culture containers at the enrichment step. However, using the MPN-mode can result in an unwieldy number of plates and slants. In addition, the challenges of preparing replicate subsamples of solid materials for inoculation also apply to this test.

Kenner and Clark (1974) proposed a multiple-tube (MPN) analysis to quantify *Salmonella* using dulcitol selenite broth for enrichment with incubation at 40°C for 1 and 2 days. Following enrichment incubation, the surface scum is removed from the sample and spread on both sections of a divided plate containing brilliant green and XLD agars. Results on brilliant green and XLD agars are as described for Standard Methods 9260D above. A second confirmation step with transfer onto Klinger iron agar (KIA) or TSI is also suggested. KIA slants are incubated overnight at 37°C. Cultures confirming to be *Salmonella* have a red slant, but have a characteristic H_2S black butt that is also acid-yellow with gas bubbles. The Kenner and Clark method can be used as an alternative to EPA Method 1682 described below.

The only method listed in the Federal Register for analysis of *Salmonella* spp. in biosolids is EPA Method 1682 (U.S. EPA, 2006b). This method is used in an MPN format where a defined amount of sample is enriched in tryptic soy broth (TSB) for 24 hours at 36°C. After incubation, the TSB enrichment is spotted onto modified semisolid Rappaport-Vassiliadis (MSRV) media. After allowing the spots to absorb for 1 hour, they are incubated at 42°C for 16-to-18 hours. Samples containing *Salmonella* will produce colonies with white halos indicating motility. Presumptive colonies are isolated on XLD, LIA, TSI, and urease test media, with associated reactions as described above. *Salmonella* spp. positive samples produce a neagative reaction in urease broth. Samples that are positive on XLD, and either LIA or TSI or both, and negative in urease broth are confirmed against polyvalent-O antiserum. A positive reaction with the antiserum is a confirmed test for *Salmonella*. Method 1682 specifies that multiple tubes are inoculated with 20 mL, 10 mL and 1.0 mL of homogenized sample, as described under section 2.1.2.1 Fecal Coliforms. These volumes represent 2.0, 1.0 and 0.1 g wet weight of original sample, respectively. To reiterate, for solids samples, accurately measuring such volumes of a heterogeneous suspension is difficult at best.

2.1.2.3 Enteric Viruses

There are more than 140 different enteric viruses known to colonize humans (AWWA, 1999). These viruses are shed in feces and include enteroviruses, rotavirus, hepatitis A and E, caliciviruses, noroviruses, adenoviruses and reoviruses. The method for measuring enteric viruses is a defined ASTM method for recovery of viruses from wastewater sludges (ASTM, 1994). Viruses are extracted and the extract can be analyzed as is, or it can be concentrated if it is suspected that the chosen viral assay method will not have enough sensitivity to detect the enteric viruses. The method to be used for viral assay is not specified, but typically involves cell culture with screening for cytopathic effects and plaque formation (Yates, 2005). Cell culture techniques can take up to 4 weeks to complete. To properly conduct cell culture analyses, highly trained staff, strict QA/QC and specialized laboratory facilities are required. Few utility laboratories can afford to maintain these capabilities, and most utilize contract laboratories to conduct such analyses.

2.1.2.4 Helminth Ova

Helminths are a group of parasitic organisms whose eggs or ova can be spread through feces of infected humans and animals. To test for viable helminth ova, a procedure developed by Yanko (1987) is applied to classify total *Ascaris*, viable *Ascaris*, *Trichuris*, *Hymenolepis*, and *Toxocara* in solid samples. The test requires lengthy processing (up to five weeks) and is very detailed and specific. Each step is subject to analyst variability. It has been noted that helminth ova tend not to be found in large numbers. Therefore, the patchiness of the occurrence of these pathogens might be missed by only periodic testing. The level of training and analysis time involved means that few utilities can conduct analyses for these organisms more than a few times a year with typical laboratory staffing levels. For more frequent monitoring, use of contract laboratories is the typical practice.

2.1.2.5 Methods Summary

Regulations pertaining to biosolids include well established testing methodologies, as summarized above. Viral and helminth assays are currently too lengthy to be conducted with any routine frequency at typical wastewater utility operated laboratories. Biosolids research regarding analytical methods has focused on the accuracy of the laboratory testing method including such issues as media selectivity and clarity of reaction (i.e. color change), and QA/QC. Typically, the 95% confidence intervals for microbial densities using EPA-approved analytical methods span an order of magnitude or more. However, the errors and variability in the data caused by sampling and handling methods may be greater than those from the laboratory methods and are mostly overlooked (Khan and Nortcliff, 1982; Lambkin, 2004). These errors can be credited to variations in sampling devices, sample volume, handling, reduction methods, population heterogeneity, and sample heterogeneity (for example, large pieces of wood in compost). Many of these sample variations and uncertainties can be exacerbated by the sampler's test day judgment. Any slight variation or contamination during the sampling and handling can result in false measurements and/or wasted lab analysis time. This type of variability is one focus of this research.

2.1.3 Biosolids Sampling and Handling

The objective of sampling is to obtain a sample or samples that are characteristic of a population - in this case biosolids products – without the need to collect and analyze the entire population. As discussed by van Elsas and Smalla (1997), designing an appropriate sampling strategy involves taking into account a number of considerations. For the sampling of biosolids, this includes the objective of the sampling, the required data precision and accuracy, and the practicality of the sampling requirements. Precision typically refers to sample variance – in other words, the closeness of repeated sample values. Accuracy is the closeness of the sample value to the true value (U.S. EPA, 2004). Practical considerations include sample locations, sample numbers, sample size, composite samples, economics, etc. (van Elsas and Smalla, 1997). The following discussion considers information from the published literature related to the sampling and handling of biosolids.

2.1.3.1 Sampling Regulatory Guidance

Specific guidance on numbers and frequency of collection of biosolids samples for microbial monitoring are governed by system size (40 CFR 503, 1993; U.S. EPA, 1994). Regarding sampling frequency, Part 503.16 requires facilities producing less than 290 metric tons/year to collect samples for microbial monitoring once per year. Facilities that produce from 290 up to 1,500 metric tons per year must collect samples four times per year. Facilities that produce from 1,500 up to 15,000 metric tons per year must sample six times per year. Finally, facilities producing more than 15,000 tons/year must monitor monthly.

2.1.3.2 Biosolids Types

The three types of biosolids that are the focus of this project are liquid, cake and compost. Liquid biosolids resulting from PFRP or PSRP typically contain up to 7% solids. Storage of liquid material can be costly and space-consuming because of the holding tank capacity required. The same applies to transportation of liquid materials. Tanker trucks or sealed containers are needed for transportation of liquid biosolids. However, liquid biosolids are predominantly water which makes for easy loading and sampling. When sampling liquid material, it is beneficial to draw from a well mixed material to ensure a good representation of the entire sample. The best way to ensure a good representative sample is to take the sample when the material is in motion (*i.e.* from a pipe outflow, or during mixing).

Cake is a semi-solid classification of biosolids that is dewatered liquid with roughly 15 to 25% solids. Cake is stored in piles or tanks. When sampling cake, it is beneficial to collect a sample that represents the entire biosolids batch. It is best to sample when the biosolids cake is well mixed and in motion. Within the text of Methods 1680, 1681, and 1682, it is recommended to sample cake from the conveyor as it is being transported from dewatering to storage or transport (U.S. EPA, 2006, 2006a and 2006b). However, if cake is stored in piles or tanks for an extended period of time, it should be re-sampled and tested prior to land application (U.S. EPA, 2000).

Compost is biosolids material that is comprised of an organic mixture of sludge and a bulking agent (to provide carbon and increase porosity and free air space) that has been subject to aerobic, biological treatment (U.S. EPA, 1999a). The process of composting is used to decompose and stabilize the organic matter through aerobic microbial action at prescribed temperatures. There are three composting methods for biosolids: windrows, aerated static piles, and in-vessel composting (U.S. EPA, 1999a). One of the major issues with sampling composted biosolids is the possible presence of large pieces of bulking agents (wood chips, leaves, or yard waste). These agents are too large for most sampling and testing apparatus but often have a large influence on both the characterization of composted biosolids and the calculations of microbial density (organisms/gram). If compost is screened to recover bulking agent, sampling after screening can help minimize this issue. Compost should be sampled for microbial content as soon after the prescribed active composting period as possible to meet PFRP and vector attraction reduction (VAR) requirements. Re-sampling of finished compost that is subject to storage in curing piles should be conducted shortly before land application.

2.1.3.3 Sampling Procedures

Methods 1680, 1681 and 1682 specify recommended sampling procedures for materials from digesters, conveyor belts, or bins, drying beds or similar locations (U.S. EPA, 2006, 2006a, and 2006b). Another good reference is the New England Interstate Water Pollution Control Commission report *The Wastewater Treatment Plant Operators Guide to Biosolids Sampling Plans* (NEIWPCC, 2006).

The EPA methods specify that sample containers must be washed, acid rinsed, wrapped in foil and autoclaved before use. When collecting a sample from a digester or storage tank, typically containing liquid, the discharge pipe must be purged of old biosolids. The sample should then be collected in a one-gallon sterile bag so that only the flow touches the inside of the bag. Approximately 0.5 inches of headspace must be left in the bag to accommodate gas production. The main point in guidance documents for biosolids sampling techniques is that it is most important to obtain a representative sample or cross-section of the biosolids flow.

When sampling biosolids from conveyor belts, typically associated with cake, a sterile scoop should be used to transfer the biosolids cake from the conveyer into a sterile container without mixing or transferring to another area of the belt. For samples in a bin or drying bed, the upper six inches of material must be moved aside. The underlying material should be divided into four quadrants. A scoop or auger should be used to take a sample from each quadrant. All four samples should be combined in one sterile container. Once all four samples are collected, the mixture should be poured onto a sterile surface and mixed back on itself several times. It is noted that curved scoops should not be used because they produce bias toward a certain sized particle. While a composite of a compost pile is described, the aerial extent of the surface cleared for sampling is not provided.

According to *Control of Pathogens and Vector Attraction in Sewage Sludge* (The White House Document; EPA/625/R-92/013), when sampling dry biosolids or those consisting of more than 40% solids, composite samples should be obtained when the biosolids are in transit on a conveyor belt, making sure to collect samples from across the width of the conveyor (U.S. EPA, 1999). If these solids are in piles, a sampling grid should be drawn making sure to weight the edges and interior appropriately. If the pile is older than 24 hours, a pile cross section should be collected as a sample. When in piles, biosolids should be sampled in proportion to the geometry of the pile. In other words, greater volume/weight of biosolids should be obtained from the tallest sections of the pile.

There are currently no required sampling standards for biosolids storage. Microbial enumeration results may vary widely with variables such as climate, age of the pile, and location in the pile from which the sample is collected. The sample being tested needs to be representative of the entire pile. Recommendations for numbers of individual samples to be used to produce composites from storage piles are provided by EPA Region VIII based on plant flow. Facilities with design flows less than 1 million gallons per day (MGD) must collect 27 discrete samples composited into one sample and analyzed once per year. For facilities with design flows between 1 and 10 MGD, 42 discrete samples to produce four composite samples quarterly must be collected. For facilities with design flows between 10 and 100 MGD, the number of discrete samples increases to 48 to produce six composite samples bi-monthly. And finally, facilities with design flows greater than 100 MGD must collect 84 discrete samples to produce 12 composite samples for analysis monthly (Brobst, 2001).

2.1.3.4 Sample Size

There is some variation in guidance materials regarding sample size. The *Plain English Guide* (EPA 832/R-93/003) recommends a sample size of 1 to 4 liters, to be split at the laboratory into approximately 300 mL samples (U.S. EPA, 1994). The *White House Document* (EPA/625/R-92/013) and the *Plain English Guide* recommend collecting samples in quart (approximately equivolent to one liter) containers or smaller (U.S. EPA, 1994 and 1999. *Preparing Sewage Sludge for Land Application or Surface Disposal* (EPA831b-93-002a) states that 1-liter samples should be collected when the sample will be analyzed for pathogens or indicators, but also recommends checking with the laboratory for the proper sample amount (U.S. EPA, 1993). The proposed Methods 1680, 1681 and 1682 recommend collection of 1-liter samples.

2.1.3.5 Sampling Containers

Samples should always be placed in sterilized, well-cleaned containers. The recommended sample containers and sampling equipment varies greatly, but all guidance documents agree that samples should be collected and stored or transported in wide-mouthed containers made of autoclavable plastic or glass. Acceptable containers would be constructed of glass, plastic or stainless steel (e.g. unused paint cans). Containers that are not recommended because of potential contamination are galvanized steel, or anything zinc-coated. The *White House Document*, the *Plain English Guide*, and EPA Methods 1680, 1681 and 1682 mention the use of pre-sterilized bags as acceptable sample containers (U.S. EPA, 1994, 2006, 2006a, and 2006b).

New guidance in EPA methods 1680, 1681, and 1682 specify sterile (plastic) bags for digester samples. Manual manipulations of plastic bags can make it difficult to assure that "only sample touches the inside of the bag." For conveyor belt or bin samples, only glass or plastic is specified (U.S. EPA, 2006, 2006a, and 2006b). Wide-mouth containers are preferable because they can be more easily filled with sample without touching the insides of the container. Glass provides a material that can be autoclaved repeatedly without deformation; however, glass containers are subject to breakage during shipping and handling.

2.1.3.6 Sampling Tools

There are various types of sampling devices appropriate for various sizes, shapes, depths of sample, and types of sample to be collected. Coliwasas are ideal for small-area liquid sludge samples and are good for single one-time-use sampling. Certain coring samplers that would be best suited for a biosolids pile that could potentially be layered, thus requiring a straight core sample. Other simpler types of samplers include pitchers, conduits, shovels and trowels, all of which should be made of a material that will not contaminate or alter the physical or chemical properties of the biosolids sample collected. Some common sampling devices and their uses are summarized in Table 2-4. These devices can be made out of metals or plastics. Ideal materials for samplers would be glass or Teflon; however, these types of samplers would typically be too fragile or prohibitively expensive. The choice of sampling device can affect the representativeness of the samples collected. If sampling devices are utilized, they should be completely decontaminated between samples and sampling events. Utilizing the exact same sampling apparatus on the same day requires cleaning and decontamination so as to not cross-contaminate grab samples.

Recommended equipment for sampling also include trowels or scoops made of stainless steel, Teflon or glass. The *POTW Sludge Sampling and Analysis Guidance Document* (EPA 833-B-89-100) states that the use of Teflon, glass, or stainless steel is preferred, but plastic, steel and aluminum are acceptable as long as the steel is not galvanized and nothing is coated with zinc (U.S. EPA, 1989). The *POTW Sludge Sampling and Analysis Guidance Document* (EPA 833-B-89-100) also mentions the use of graduated glass or plastic cylinders for grab and composted samples, and stainless steel pitchers for samples taken from taps (U.S. EPA, 1989). For mixing the samples before splitting them, the same guidance document recommends the use of glass or Teflon-coated stirring rods or stainless steel spoons (U.S. EPA, 1989). The *White House Document* recommends using a thief sampler if sampling from a tank, and augers or shovels for obtaining samples from a pile (U.S. EPA, 1999).

Table 2-4. A Summary of Common Sampling Devices.

Device	Sample size	Biosolids physical characteristics	Device size
Composite liquid biosolids sampler	Composite	free-flowing liquids	Typically 5 up to 15'
Thief sampler	Composite	granulated or powders	2-4'
Trier sampler	Composite	semi-solids	up to 4'
Auger	Cross section or discrete	hard packed solids	variable
Shovel and scoops	Discrete	granulated, powders, or loose solids	variable

After Table 2.4-4, Biosolids Reference Sheet, U.S. EPA website.

Appropriate sampling and handling practices and guidance can be informed by considering lessons learned from the hazardous and solid waste fields as they must also collect representative samples of heterogeneous materials including liquids, sludges, semi-solids, and solids. Tinklenberg and Smith (1990) investigated the use of two relatively low-cost, sampling devices for obtaining hazardous waste samples to meet quality control requirements for performing TCLP analyses. The two devices tested were the thief sampler (approximately $200) and the trier sampler (approximately $50). The thief sampler works well with relatively dry material and is comprised of a tube within a tube each with holes on one side. To collect the sample the tubes are rotated so the holes are not lined up and a solid shaft results. The sampler is inserted into the pile and withdrawn. The tubes are then rotated so that the holes line up, so that sample can be collected, the thief withdrawn and then emptied into a container. The tier sampler is more appropriate than the thief sampler for moist or densely packed material. It consists of a half tube that can be inserted into a pile to acquire a core sample by rotating and withdrawing the sampler. The length of these types of samplers can be up to 5', but are more typically 3-4'. Typical compost piles can be approximately 6-10' high and 15-25' across. Thus, using these types of sample devices to collect a core spanning the entire depth of a pile can sometimes be challenging. The diameter of these types of sampling devices may bias the distribution of particle sizes collected (towards the smaller end) if the material being sampled is heterogeneous with respect to mean particle size.

For solid waste, increasing the numbers and randomized locations of sample increments that comprise a composite sample can increase the representativeness of the final sample analyzed (U.S. EPA, 2004). Statistical analyses can be used to account for variability in samples resulting from heterogeneity of the materials, and for errors resulting from contamination and interaction with samples and containers. Studies conducted in the solid waste field indicate that the variance in analysis results from previous sampling events should be used to determine the number of samples to be collected in the future. Processes/facilities with a history of higher variability should be sampled more frequently, with a greater number of samples for compositing. In one study, a minimum of 12 and a maximum of 30 discrete soil samples were recommended to comprise a composite based on the number of cubic meters of materials being sampled (CEN, 2000). Lambkin et al. (2004) reported that in a study of compost materials, a composite sample of no fewer than 15 discrete samples was needed to provide 80% confidence that the measurement was representative of the true population value. Once the number of discrete samples is determined statistically, a random-number block method can be used to determine locations of the discrete samples across the aerial extent of the pile or tank (U.S. EPA, 2004). In another study, discrete sample increments spanning 200-300 m^3 of compost was proposed (Laraia et al., 2002).

Lambkin et al. (2004) investigated the importance of numbers of samples on precision in sampling sludges, biowastes (the organic fraction of municipal solid waste) and treated soils to meet regulatory requirements. In sampling centrifuged, dewatered cake, the researchers compared four sample collection methods. Those methods included: 1) collecting nine samples from the centrifuge output; 2) collecting 14 samples from a cake heap using a trowel at 1-25 cm; 3) collecting seven samples at 25-50 cm from the side of the heap using a corer; and 4) using an excavator to cone and quarter the entire pile, then manually collecting four samples. All samples were composited separately for each collection method. The coefficient of variation within each sampling method was less than 6% for metals analyses. There were no significant differences at the 95% confidence level in the mean metals content of the samples. However, microorganisms in biosolids, the concept that microorganisms are discrete particles and inherently may not be as evenly distributed through a sample in a similar manner as a dissolved chemical constituent, must be recognized. The results of this study suggests that multiple grab samples be tested for microbial content to capture variability, and that well mixed composites might adequately represent the mean.

2.1.3.7 Storage and Shipping

After collecting samples of biosolids, there is a brief period of time during which testing should be completed. The sooner the sample is tested, the more accurate the results. For example, the initial enrichment inoculations for *Salmonella* testing should preferably be completed within two hours of sample collection (U.S. EPA, 2006b). Extended storage times can only cause larger bases for error. Storage conditions can also affect regrowth and decay of microorganisms within biosolids samples. Guidance in *Standard Methods* and proposed EPA method 1680 suggest that samples be stored in coolers or refrigerators at temperatures maintained below 10°C for no more than six hours unless otherwise specified in the 503 regulations for fecal coliforms and *Salmonella* (APHA et al., 2005; U.S. EPA, 2006). Class A compost, and Class B aerobically and anaerobically digested matrices may be analyzed after 24 hour holding times when analyzed using EPA Methods 1680 and 1681 (U.S. EPA, 2006c). Class A matrices may not be analyzed beyond 6 hour holding times for *Salmonella* (U.S. EPA, 2006c).

There is some variation among existing guidance regarding the maximum length of time samples may be stored or in transit to a laboratory before analysis, and how to cool the sample during storage and shipping. For cooling, the *Plain English Guide*, the *White House Document*, *Preparing Sewage Sludge for Land Application or Surface Disposal*, and *Wastewater Sampling for Process and Quality Control* (MOP-OM-1) all recommend a cooler filled with a slurry of water and ice or bagged ice (U.S. EPA, 1993, 1994, and 1999; WEF, 1996). The *Plain English Guide* states that the use of blue packs is also acceptable, and the *White House Document* states that coolers with blue packs and a temperature of 0°C should be used for shipping. The Water Environment Foundation (WEF) MOP-OM-1 manual states that refrigeration at 4°C is also acceptable. Of the other guidance documents reviewed that discuss cooling, the *NJDEP Field Sampling Guide* simply states that the sample should be chilled to 4ºC (NJDEP, 2005), and the *POTW Sludge Sampling and Analysis Guidance Document* (U.S. EPA, 1989), mentions the use of an "iced cooler" for storage, but the use of blue packs and insulation during shipping. Table 2-5 summarizes the allowed holding times and temperatures during storage and shipping from each of these guidance documents.

Table 2-5. Allowed Holding Times and Storage/Shipping Temperatures.

Guidance Document	Holding Time	Temperature
White House Document		
Fecal Coliform *Salmonella*	24 hours	4°C
Enteric Viruses	up to two weeks	-18°C
Helminth	one month	4°C
For all analyses:	within one hour 1-6 hours 6-24 hours	No Cooling Required ≤10°C 4°C
Plain English Guide		
Fecal Coliform *Salmonella*	24 hours	4°C
Enteric Viruses	up to two weeks	0°C
Helminth	one month	4°C
For all analyses:	1-6 hours	<10°C
EPA 831B-93-002a		
Fecal Coliform *Salmonella*	24 hours	4°C
Enteric Viruses	up to two weeks	frozen
Helminth	one month	4°C
NJDEP Field Sampling Guide*		
Fecal and Total Coliform	6 hours	4°C
Enteric Viruses	8 hours	4°C
Salmonella Helminth	24 hours	4°C
EPA 833-B-89-100		
For all analyses	6 hours	<10°C, 4°C during shipping
MOP-OM-1		
For all analyses	6 hours	4°C

*guidance specifically applies to liquid matrices

It should be noted that for the Part 136 methods, the holding time is either less or the same as the *White House Document*, but with slightly different temperatures (U.S. EPA, 2006c). Specifically, if using 1680 or 1681 for fecal coliforms, and assaying Class A composted or Class B aerobically or anaerobically digested materials, the holding time is 24 hours. All other combinations for the three EPA methods (1680/1681, 1682) are 6 hours. Also, the holding temp is <10ºC rather than frozen for all three EPA methods for all sample types.

2.1.4 Sampling Information from Other Disciplines

As stated above, many other disciplines involve sampling and analysis of heterogeneous materials. Some of these disciplines are more mature than the biosolids field, and research findings from these areas might provide some insight into developing biosolids sampling and handling guidance. These fields include the hazardous waste, solid waste, mining, and soils disciplines.

When sampling hazardous waste piles, it is essential to collect composite samples to represent the entire pile area without testing every cubic inch. Parkhurst (1984) conducted a study that demonstrated that by dividing the sampling area into triangular quadrants, as opposed to square grids, one will not only cut costs but will also obtain more representative samples. However, their application was in searching for toxic materials buried in abandoned dumps and landfills. The final objective in sampling hazardous waste sites is often not to obtain an unbiased concentration of waste measurement, but rather to locate pockets of waste. Nevertheless, some information from the hazardous waste discipline can be helpful in designing biosolids sampling schemes. Parkhurst (1984) also discusses choosing between random or regular sampling across a grid. If choosing a randomizing scheme, there is no advantage to using a triangular grid; a square or rectangular grid would serve just as well. It can also be concluded that if variations in microbial content within a biosolids pile is random, then a regularly spaced grid with regular sample locations, which is easier to apply, would work as well as a randomized sampling scheme.

From the hazardous waste discipline, sampling and testing of soil stockpiles to demonstrate levels of contamination/clean-up are also common. When sampling soil stockpiles, the goal is to obtain a single analytical result from a heterogeneous mixed pile of soil. Lamé et al. (2005) describes a strategy adopted in the Netherlands aimed at obtaining representative results from samples of heterogeneous soil stockpiles. They recommended combining a large number of random increments to capture pile heterogeneity, with increment samples collected from the outside edges to the interior core. The numerous samples would then be combined to produce a well-mixed composite sample. The group concluded that for a stockpile, two composite samples of 50 increments each should be collected and analyzed (Lamé, 2004). Those two composite results were then compared to check the heterogeneity within the stockpile. This approach was validated on 2,570 stockpiles across the Netherlands.

From the mining discipline, a considerable amount of information regarding sampling and sampling devices has been published (Reiners and Mohrhauer, 1994; Mathar and Rauhut, 1983). Reiners and Mohrhauer (1994) reviewed sampling devices for collecting samples of ore with the goal of providing accurate and reliable standards for the industry. Most devices were functionally able to obtain samples along a variety of conveyor devices, and also grind or mill materials to an acceptable particle size. Hammer devices were suggested as preferred devices, but statistical analysis of accuracy and representativeness of samples was not explicitly discussed. Mathar and Rauhut (1983) suggest the use of a stochastic model incorporating an estimated population mean and standard deviation for the parameter being measured from past testing to determine the weight of material to be sampled. Guidance for sample collection from the solid and hazardous waste disciplines also suggests that numbers and sizes of samples be based on prior analytical results (U.S. EPA, 2004).

2.1.5 Research Questions

Current guidance reviewed during the literature review contains details that specify the sample numbers, sample sizes, sampling scheduling, sample handling and sample analyses, but offers little detail as to sample techniques, sample locations within tanks and piles, or how to obtain statistically representative samples (U.S. EPA, 1999). A suite of newly revised EPA methods (1680/1681 and 1682) attempts to address these issues by describing sampling from digesters, conveyor belts, and bins and drying beds (U.S. EPA, 2006, 2006a, and 2006b).

This project examined the current practices for obtaining sludge and biosolids samples for regulatory compliance with microbial testing, and reviewed methods for obtaining representative samples. The representativeness of sample composites, sampling frequency, and issues dealing with sample heterogenity were examined. The reproducibility of homogenates obtained using specific procedures provided in EPA guidance was also examined.

To obtain the most representative biosolids samples, some key questions possibly include:

- How many increments are needed to produce composite samples?
- For biosolids stored in piles, what methods should be used to generate grid patterns?
- What are the best sampling methods to assure that samplers do not microbially compromise/contaminate samples during sampling?
- For samples with large variations in particle size distribution - typically compost - how should variation in particle size for inoculation of bottles and test tubes be accomplished to account for larger particles?
- What statistical tools are practical for determining increment number and sizes required to obtain a representative sample?

2.2 Utility Survey

The literature review did not fully uncover important information concerning how biosolids sampling is actually conducted in practice, why it is conducted in specific ways, accountability for sampling procedures, and ultimately, what factors influence procedures across a range of facility characteristics. The project team's approach to better addressing these information needs was to apply sound survey techniques (i.e. conduct a utility survey).

The survey provided up-to-date feedback from biosolids sampling stakeholders. Even if the technical methodology for sampling is currently sufficient, or if new methods are proposed as more technically sound, effective implementation may hinge on non-technical factors. It is important to understand how decisions and practices specific to sampling biosolids materials and products are determined, managed, implemented and assured. Furthermore, the survey tool helped to assess variability across utilities by size, region, management structure and training requirements.

The primary objectives of the survey included:

- identifying existing biosolids sampling/handling techniques;
- identifying limitations of current techniques and other concerns/problems identified by samplers;
- identifying existing guidance that may inform this work;
- identifying potential Phase II and III utility participants;
- identifying where sampling of final biosolids materials is occurring in order to target guidance to the right people; and
- providing information for developing the communications plan.

In developing the survey tools and target mailing list, a number of factors were taken into consideration. These factors included ensuring that the survey was defensible with respect to representativeness (geographical, process, size) and bias, and whether survey results could be matched with treatment processes and biosolids products.

A two-part survey tool was developed and hardcopies were mailed to selected utilities. The first part targeted the managers of utilities and biosolids product producers. The second targeted individuals at the facility who actually sample biosolids intended for microbial analyses. The two surveys were linked using a randomly-assigned facility-specific number, so that manager and sampler surveys could be compared. Copies of both surveys are contained in Appendix A to this report. The objectives of the manager survey included:

- identifying basic utility/WWTP and biosolids processing characteristics:
    - to guide proper follow-up;
    - to adequately characterize the sample population and to address possible future questions of non-representativeness and/or bias; and
    - to match specific sampling information garnered from this and the sampler survey with important facility characteristics, including biosolids processing methods and product types.
- identifying utility biosolids samplers and/or contract processors for follow-up surveys;
- identifying guidance/resources that management provides to field samplers;
- identifying utility concerns to help define "the problem" and to help refine project goals and target communications and guidance;
- encouraging managers to motivate samplers to complete follow-up survey;
- identifying potential Phase II participants; and
- identifying where sampling of final biosolids materials is occurring and by whom.

The instructions explained that more specific questions about biosolids sampling were included in the survey to samplers, and that the information from the manager would help us ask questions that are appropriate to the processes used at that utility.

The aims of the sampler survey were different from, and complementary to, the information requested from the managers. The objectives of the sampler survey included:

- identifying guidance/resources used to guide field sampling;
- identifying specific biosolids sampling procedures utilized; and
- identifying sampler concerns.

The surveys were sent to 266 utilities representing a distribution of facility sizes, based on design flow, and within each EPA administrative region.

2.2.1 Synopsis of Utilities Survey Results

2.2.1.1 Overview

A valuable and important part of the data gathering and information retrieval process for this project involved the use of survey methodology. Hardcopy surveys were sent to more than 200 utilities. The surveys targeted both facility managers and parties responsible for conducting onsite sampling of biosolids for the purpose of microbial analysis. The surveys were designed to provide both technical and non-technical information useful in determining the range of current sampling procedures and information exchange practices nationwide, and to identify ways to improve these activities.

The primary focus of the ensuing survey analysis was on non-technical responses to assess the nature and source of training, education, and communication-related activities. This information was used to identify how these aspects of utility operations could be enhanced. A general quantitative review of the survey responses and a qualitative evaluation of facility sampling and management practices were also conducted. In conducting qualitative analysis of the survey results, responses to specific open-ended questions and for those responses that indicated that outside vendors were used for final biosolids processing were closely reviewed for a subset of the surveys. In particular, surveys were selected based on the thoroughness of responses and articulation of useful information. This resulted in careful review of approximately 17% of the surveys received.

2.2.1.2 General Survey Response Data Summary

Of the 266 pairs of surveys mailed, 69 manager and 51 sampler surveys were returned. Of the 69 manager surveys, 68 contained answers to most, if not all, questions. Of the 51 sampler surveys returned, 46 contained answers to most if not all questions posed. For those questions where responses were quantitatively analyzed, the calculated percentages for the respondent answers are summarized below.

The utilities responding to the surveys dispose of their sludge and/or biosolids in a number of ways. Based on the responses, 74% of the facilities perform final processing and 26% contract final processing with an outside vendor. When noted that all processing was done in-house, respondents were sending at least part of their finished biosolids for commercial uses (primarily landscaping and agriculture applications). Of those processing all material in-house, 46% indicated at least part of their end product went to commercial use in comparison to 25% who contract out final processing. Responses concerning product classification revealed that 30% of the facilities are producing Class A material and 58% are producing Class B, whereas 22% produce material with no classification.

When asked whether microbial analyses are conducted on or off-site, responses were mixed and in some cases inconsistent within the survey questionnaire. Only 30% of the facilities that initially indicated they did not conduct onsite microbial analysis elsewhere in the survey also stated that analysis was conducted off-site. In some instances certain analysis activities were end-use specific or were, in part, conducted onsite for quality assurance purposes, but all final analysis was conducted off-site. Approximately 37% of respondents indicated that microbial analyses were done onsite. Approximately 25% of the respondents indicated they did not conduct microbial analysis either on or off-site. Of this group, most indicated their material was either incinerated or landfilled. Four respondents indicated that non-analyzed processed material was used commercially with three of those four indicating it was for agricultural purposes; the other provided no further details.

Of the respondents indicating that they do not conduct microbial analysis, conflicting responses about analysis were observed in several instances. This appeared to be due in part to having some material tested in-house for purposes other than microbial testing or specific to a certain pathogen in advance of disposal or off-site transport. Therefore, it is possible that some level of analysis is occurring for much of the classified product, but survey participants were not clear about responsibility or procedure.

2.2.1.3 Sample Collection Responsibility

Of the *managers* responding to the surveys, 58% indicated the person responsible for sampling was a plant superintendent or operator. Of the *samplers* responding, 50% indicated either a plant superintendent or plant operator conducted sampling, 46% indicated other utility employees, and 4% an independent contractor. Sampler respondents indicated more than one employee was involved in sampling activity at a given plant (answers commonly ranged from one to 14, with a mean number of 6.4). For those plants where the respondent was a superintendent or operator, the mean number of samplers was eight.

These results suggest two possibilities. One is that there are potentially a number of individuals taking samples at each plant, including many plant technicians. Second, the utility managers appear to have frequently given the sampler surveys to their most experienced and/or knowledgeable employees to complete. The survey responses from samplers, therefore, may be overstating the experience and knowledge of the average sampler. The practical implication is that protocols, instructions and communications strategies must be appropriate for a diverse audience, and yet perhaps give special attention to plant technicians.

2.2.1.4 Certification and Training

Although the responses between managers and samplers at the same utility concerning certification requirements were not always consistent, 39% of all respondents indicated they were certified for sampling or field-testing. Specific to manager survey responses, 51% indicated their utility required certification for employees who collect samples. Interestingly, the certification rate reached 94% for those respondents indicating that their state requires certification, whereas none of the respondents indicated they were certified in those states where it is not required.

Questions pertaining to certification revealed that often the "requirement" was for *operators,* and not specifically for samplers or sampling activities. In several instances, managers indicated certification was required, but the sampler response from the same facility indicated otherwise. In some instances the manager implied certification was required, but the sampler's response clarified this certification was for plant operators or supervisors only.

As previously mentioned, it appears certification steps are taken predominately by utilities within states that require or offer some form of a certification process. Typically, the certification is for plant or facility operations and oversight, rather than specifically for sampling. Some examples of specific certifications mentioned include:

- Certification of Labs and Environmental Measurement, NJAC 7:18
- Texas Wastewater 'C' Certification
- Operator Class III WWT

The responses to questions related to aspects of sampler training and guidance revealed a great deal of variability across utilities. Of those samplers responding to the survey, 78% learned from a demonstration by another utility employee before they began collecting samples for microbial analysis on their own, but only 51% indicated that they had collected samples with any supervision before soloing. Twenty percent of the sampler responses indicated that they neither observed sampling procedures nor were observed conducting this task before soloing, and 11% said that they had received no hours of specific training for sampling of biosolids for microbial analysis.

For those samplers that are trained, there appears to be high variability in the level of training and oversight samplers receive. The responses concerning training indicate that demonstration by a supervisor, operator or other employee is commonly used. Sampler responses about the number of times being supervised typically ranged from one time up to 10 times, and usually were similar to the numbers provided by their facility manager. However, in one instance the manager indicated supervision 10 times and the sampler said from 90-100 times. Workshops, though not frequently mentioned, were a source of training for some respondents and typically were conducted by an agency or association. Those respondents that received video training commented on the effectiveness of this approach. Several that did not receive such training suggested using videos as a training technique. Very few used charts or pocket guides as a training tool. Written guidance that was often mentioned was in the form of internal SOPs or agency guidance documents and fact sheets given to employees responsible for sampling.

Despite many responses indicating limited structured or focused training for microbial analysis sampling, respondents overall commented that they felt confident about the training that they received prior to collecting their first samples. Training was rated good or excellent by 73% of the respondents (with 40% stating excellent). Checking excellent on the survey form reflected the respondent agreeing with the statement: "I understood the importance of proper sample collection and handling, and I knew exactly what to do." The high level of confidence in training prevailed even among respondents who indicated they had received little or no training.

- 64% of the samplers who never collected a supervised sample before sampling on their own believed they received "good" or "excellent" training.
- 78% of the samplers who never collected a supervised sample nor observed another employee take a sample rated their training as "good" or "excellent."
- 75% of the samplers who reported zero hours of training prior to taking their first sample rated their training as "good" or "excellent."

The responses from certified personnel reveal an even higher level of confidence about their training: 88% indicated their training was good or excellent compared to 56% of non-certified respondents. There was also variability in confidence relative to the type of material sampled. Despite the difficulty of sampling from piles of cake or composted material, samplers of these materials expressed greater confidence in their training than respondents who sampled from liquid biosolids (73% good or excellent for the drier materials versus 63% for liquid).

2.2.1.5 Utilization of Guidance Documents and QA/QC Programs

Again, there appears to be wide variability in the use of specific guidance documents and/or quality assurance/quality control programs among the utilities. There is also evidence of disparity in the responses about the use of guidance documents between managers and samplers, in some instances at the same facility. General quantitative results concerning use of a QA/QC guidance document for sampling were as follows:

Manager survey:

- 56% say yes
- 44% say no

Sampler survey:

- 67% say yes
- 33% say no

When asked "are microbial sampling practices included in regular review for people who collect samples?" managers were more likely to respond affirmatively.

Manager survey:
- 52% say yes
- 48% say no

Sampler survey:
- 36 % say yes
- 64 % say no

The responses concerning QA/QC programs revealed that sampling technique and approach is seldom a focal point in the QA plan. However, a few utilities did comment on details within their program specific to sampling procedures. For example one manager's response stated they develop "SOPs to ensure samples are representative and uniform in collection; internal plan with chain-of-custody, revision dates, and hands-on training." Another respondent stated their QA/QC objective is "to meet state marketing and distribution permit limits and to have every batch sample collected for bacteria, nutrients, and metals with all data recorded in B.D.M.S. program" (biosolids data management system). Less than half of the respondents indicating the use of a QA/QC program were applying quality controls that were specific to a particular product end use. Only one respondent mentioned use and implementation of an environmental management system (EMS) at the facility. Of 11 responses mentioning the use of an outside vendor, five indicated they had a QA/QC program in place and only one mentioned that the vendor required a QA/QC plan. Certification was required by five of these utilities.

When respondents indicated the use of guidance documents, there were only a few different "formal" documents mentioned including federal and state specific resources. Examples of guidance documents referenced include:

- EPA 831B-93-002a
- EPA/625/R-92/013 (revised July 2003)
- White House Document *Environmental Regulations and Technology Control of Pathogens and Vector Attraction in Sewage Sludge*
- EPA *Plain English Guide*
- WEF MOP OM-1
- 40 CFR Part 503 regulation
- NPDES Permit guidelines
- Certification of Labs and Environmental Measurement, NJAC 7:18
- *NJDEP Sampling Guide*
- Texas Education Extension Service training for equipment manufacturer representatives and workshops by TCEQ

Several utilities developed internal SOPs and, within those documents, outlined sampling protocols to be applied. In some instances, the QA/QC program includes specific sections targeting sampling activities. It appears that many of the utilities rely on laboratories to provide some form of guidance on sampling methods, and they are usually the source of both the information about sampling and the actual specified containers for sample handling.

2.2.1.6 Issues of Concern

Few respondents articulated thoughts or suggestions about issues or concerns regarding biosolids sampling for microbial analysis. However, the respondents who did address this question provided good suggestions. Several suggested that having a new or improved written plan in place for employees would be helpful, and that using a video for training would be a good practice. They suggested retraining at least once per year. Some indicated they were concerned about the health risks of direct contact with sludge and biosolids, exposure to airborne pathogens, and to viruses that may be contacted during sampling. The following are examples of other comments:

- "Is there a better indicator of pathogen presence than fecal coliform or *Salmonella*?"
- "After initial review of sampling procedures, we should have pre-trained individuals oversee trainees until such time trainees can sample proficiently."
- "Operators don't seem as familiar with Standard Methods as lab personnel."
- "What volume of biosolids does each individual sample reliably represent?"
- "It would be beneficial for all parties if guidance manual were established, and when followed would ensure compliance with all agency requirements."
- "Would like to be better informed about holding times if applicable, preservation practices, difference in use of plastic versus glass, and what parameters are being tested and why."
- "A manual from an agency would be beneficial."
- "Lengthen on-the-job training."
- "An SOP for sampling - whether biosolids are liquid, cake, or compost - would be helpful. This survey pointed out that an SOP in written form is needed. While the method being used is a common sense and logical approach that is okay, it should be in written form."

2.2.1.7 Sampling Practices

Answers to questions about actual sampling practices varied widely as did batch sizes and sampling quantities. Several respondents indicated they were following EPA guidelines requiring the collection of seven samples for conducting fecal coliform analysis (the guidance for seven samples comes from the Part 503 Regulations related to Class B biosolids). For example, one manager wrote, "we take one sample every 10 minutes to get seven samples required, over four weeks for fecal analysis." Another manager said that they send material (cake) to a separate study facility and take grab samples every five minutes until the required seven samples are obtained. Samplers of cake and/or compost often used sterile gloves and ziplock bags to grab samples; other utilities were using sterile spoons, scoops or buckets to obtain samples. The following are a few examples of informative responses:

- Survey #172M – Liquid volume (2 million gallons); manual hand pump used; four different locations in the lagoon
- Survey #126S – Cake (200-1200 yd^3), garden spade used, 2x/yr and shipped to outside lab
- Survey #107S – Cake up to 60 million gallons air-dried, high-density scoop
- Survey #163S – Compost (250 yd^3); shovel, sterile scoop, sterile sample bag used
- Survey #184M – Cake (scoop every two hours from conveyor to 100-ton hopper, 12/day)
- Survey #22M - Additional treatment after leaves the facility; 20x20 ft sludge pad with no preparatory steps, seven spoons per bottle collection; liquid sample bottles 1/tank, from digesters with no preparatory steps; compost samples taken from biosolids pile, different areas according to 503 requirements.
- Survey #83S – Cake from 250/min, every three hours grab samples in 60 ml cups; liquid every 25 minutes

2.2.2 Additional Synthesis of Survey Results

To support the goals of the project, additional synthesis of the survey results focused on several individual questions as well as potentially useful correlations of answers to selected questions within both the manager and sampler surveys. The overall results of the surveys demonstrated a wide variation in actual approaches to, and the activities associated with, sampling for microbial analysis along with a range of different biosolids-handling procedures among the utilities. Questions relating specifically to biosolids processes, product type and end use and associated specific sampling information were thus closely reviewed. Because not all of the surveys returned were completed with clear information or substantial content, those surveys that were completed as such were selected for subsequent quantitative analysis.

Of the completed surveys, approximately 41% of the respondents indicated that microbial analysis was conducted at the utility, whereas 43% indicated the samples were shipped to an outside lab.

There were 35 sampler responses that provided information about the biosolids process used, including at what point during the process samples were collected. Of this group, 58 total process questions were answered indicating at what point samples were collected. Out of these responses:

- 43 took samples at the end of the process
- 7 during the process
- 4 before the process
- 2 during and at the end of the process
- 2 before, during and at the end of the process

The processes applied and the number of times mentioned were:

- 11 centrifuge (dewatering)
- 7 anaerobic digestion
- 5 belt filter press
- 4 gravity belt thickening
- 4 aerobic digestion
- 3 composting
- 2 lime stabilization
- Others included miscellaneous forms of drying, storage, etc.

Those facilities that conducted sampling *before the process* included two centrifuge, one lime stabilization, and one gravity belt process. Those sampling *during the process* included four digestion processes, two gravity belt and one belt press process. Processes with sampling *during and at the end* were digestion and centrifuge dewatering, and those sampling at *all three intervals* were both centrifuge processes. Some survey respondents listed processes but were unresponsive with regards to sampling activity, other survey respondents listed processes, but it was clear that they were not collecting samples for analysis.

When evaluating potential correlation between class of material produced and end use of the product with actual sampling practices (particularly batch sizes from which samples are taken, the frequency of sampling, and handling of samples) the analysis revealed substantial variability across responses. There were 31 facilities surveyed where both the manager and the sampler responded to the surveys and provided adequate information for this quantitative assessment. Some of the respondents indicated more than one end use, and some did not indicate any end use; therefore, the number of responses does not necessarily correspond to the number of facilities responding. Of these 31 facilities, the following summarizes class and product end use(s):

Seven Class A producers with end uses

- 7 commercial agriculture
- 5 residential or commercial landscaping
- 3 vendors

Nineteen Class B producers with end uses

- 8 commercial agriculture
- 3 surface disposal
- 3 landfill
- 1 each of residential, subsurface injection, and vendor

Five Class A and B producers with end uses

- 2 commercial agriculture and commercial landscaping
- 2 landfill
- 1 residential
- 1 vendor

Cumulatively, the most frequently reported end use was commercial agriculture (17) followed by commercial or residential landscaping (8), landfill or surface disposal (5 and 3), and vendor (5).

It would be difficult to report a common response to average batch size. This was expected because one important element of the survey design was to query utilities across facility size. To demonstrate batch size variability, for the Class A facilities, batch sizes for cake material ranged from 10 yd^3 to 500-600 ton, and for compost 25 yd^3 to 800 yd^3. For Class B facilities, the range for cake was a 15-yard truck to 1200 yd^3 and for liquid from 350,000 gallons to 3 million gallons. Sampling frequency for Class A material ranged from 10g/2 hrs 12 times per day to 7/batch during 14 days for cake, with limited information supplied for liquid or compost. For Class B material, sampling frequency ranged from 6/batch every two hours to 7/batch quarterly for cake; for liquid, 1-6/batch every few hours, to 7/batch every two months, to one grab sample every 10 months.

There appeared to be little difference in methods of sampling applied for Class A versus Class B materials, with the exception that sterilization procedures were mentioned in more detail where respondents produced Class A product. Most used sterilized scoops, buckets, shovels or a spatula and placing samples in separate, sterilized plastic or glass jars or in plastic, ziplock bags. Some commented on the use of deionized (DI) water or lab-grade detergents for rinsing. The Class B responses most often mentioned ziplock bags, but also mentioned the use of sterilized bottles (plastic or glass) and, occasionally, a sterilized scoop or autoclaved bottles. Glass or plastic container sizes were also highly variable among respondents. It does appear that most handlers understand the importance of refrigeration prior to shipment. Samples were being shipped at variable times, ranging from immediately to longer than one week. Approximately 92% of all respondents answering this question shipped within 24 hours, whereas 8% held samples up to seven days.

A synopsis of the responses regarding sample containers and containment practices again reveals much variability among respondents. Plastic containers were indicated more frequently than glass, with references to PPE, HPPE, and HDPE most common. Several respondents mentioned using Nalgene containers, with a few mentioning the use of Teflon for collecting liquid samples. Most respondents indicated that they use opaque containers. Sample container sizes also are highly variable. However, however many respondents indicated using containers of 125 mL to 500 mL. Responses concerning headspace were also highly variable.

Of the respondents to the sampler surveys that expressed issues of concern or posed questions, some representative examples include:

- What are the health hazards of direct contact with biosolids?
- What are specific holding times if applicable, preservation techniques if any, the differences between using glass and plastic, and what parameters are being tested and why?
- Details about chain-of-custody requirements (some respondents indicated they didn't know what chain-of-custody requirements meant).
- What is the proper location to sample for compliance with permit/regulations?
- How do we get reasonable feedback on results confirming appropriateness of correct sample location?
- Are more in-depth soil/biosolids sample procedures available?
- Need guidance on sampling batches in situations where disposal is seasonal.
- Need more definitive universal information on methodology of collecting truly representative biosolids samples from large piles of aged, granular materials.

There were several survey responses that contained more definitive and complete descriptions of methodology and/or practices at the utility. These survey numbers are listed below and may provide additional reference information useful to this project.

- Manager surveys {#94, #99, #102, #103, #107, #148, #202, #204}
- Sampler surveys {#58, #70, #90, #91, #103, #107, #140, #148, #170, #198}

Researchers evaluated the relationships between the final disposition or end use of processed biosolids, the class of the material produced, and the presence of a corresponding quality assurance program. This information could help determine if the class and end use may have influenced decisions to create QA/QC plans, or if the presence of a plan influenced the sampling procedures.

To conduct the analysis, three categories were determined based on final disposition or product use. Each category reflects a relative level of potential risk for human exposure to pathogens.

- Higher risk – respondents sending at least part of their biosolids to residential land application or commercial land application (landscaping or agriculture). These end uses are grouped because they present the greatest concern for potential public exposure to pathogens.

- Lower risk – respondents sending their sludge and/or biosolids exclusively to landfills, incinerators, or outside vendors for further processing. These end uses are presumed to present the lowest risk for public exposure directly from the utilities, assuming the landfill operators, incinerator operators and vendors would have responsibility for safe handling.
- Other – respondents sending their waste to non-beneficial land application or "other" forms of disposal.

For this analysis, the respondents were grouped by class of waste where:

- Class A includes all respondents who at least partially classified product as Class A (some of these respondents also produce class B material);
- Class B includes respondents who list production of Class B material but may also produce some Class A material;
- no classification includes those respondents who do not list any form of classification.

Respondents were also grouped based on their use of a quality assurance program for biosolids pathogen sampling:

- Has a QA program for pathogen sampling, either written or unwritten
- Has no QA program for pathogen sampling

Table 2-6 shows the comparison of biosolids classification with QA programs. Utilities with Class A permits are most likely to have a QA program of some type. None of the respondents without a classification had a QA program.

Table 2-6. Comparison of Class with QA Programs.

	Class A	Class B	Not Classified
Has QA program	68%	41%	0%
Does not have QA program	32%	59%	100%

Table 2-7 compares final disposition or end use with the presence of a QA program. The utilities that utilize landfilling, incineration, or send their sludge and/or biosolids for further processing were least likely to have a QA program. For utilities that send their biosolids for agriculture and landscaping land application uses, fewer than 50% of respondents indicated they had a QA program.

Table 2-7. Comparison of Final Disposition with QA Programs.

	Final Disposition		
	Ag/Landscape	Landfill/Incinerate/ Further processing	Other
Has QA program	47%	21%	57%
Does not have QA program	53%	79%	43%

Table 2-8 summarizes the final disposition or end use according to the process classification in place at the utility. Of the respondents producing unclassified material, 87% send their sludge and/or biosolids to landfilling, incineration, or further processing. Many utilities with Class A material sent their biosolids to multiple uses (agriculture and landscaping), making their total exceed 100%.

Table 2-8. Comparison of Product Classification with Final Disposition.

	Final Disposition		
	Ag/Landscape	Landfill/Incinerate/ Further Process	Other
Class A	70%	60%	5%
Class B	50%	32%	16%
Not Classified	7%	87%	7%

The results above indicate that it is likely that managers and/or samplers at utilities have a higher level of awareness about the importance of both sampling for microbial analysis and having a sampling plan in place to guide the process than they do about the quality of biosolids they produce and market they serve. It is evident that, despite the disposition or end use market served, the implementation of QA/QC plans guiding sampling for safe use or disposal is limited across utilities. This could be a result of greater reliance on pre-existing "guidance" documents available to the utility. However, based on the responses concerning training and actual sampling activities, it appears quality assurance is not necessarily being prioritized.

2.2.3 Overview of Project Guidance Gained from the Utility Survey

- Collectively, the responses reveal wide variation across utilities in the approach to, and activities associated with, sampling for microbial analysis. Those utilities in states having more stringent or clear guidelines and oversight in place have more concise guidelines and consistent practices compared to the other utilities. Regardless, it is apparent that simplification in sampling activities is, for the most part, a goal.
- Laboratories are often primary sources of information specific to sampling and can be very influential in how utilities proceed with their sampling practices. There exist influences from

contracted labs. The geographic proximity of the laboratory often influences whether or not analysis is contracted out.

- There is high variability in batch sizes from which samples are collected and the frequency of sample collection within and across material types. There appears to be little, if any, real differences in sampling methods between Class A and Class B biosolids procedures.
- Dewatering and some form of digestion are the most common processing practices mentioned by respondents. Most samplers are collecting samples at the end of processes.
- Plastic containers are used more often than glass, and plastic ziplock bags are used frequently for sample handling (and the latter sometimes as the sampling tool). Tools used vary somewhat, but what is most evident is that the tools selected and sampling practiced reflects simplicity and minimal financial investment. Most seem to understand cooling/refrigeration requirements, though it appears there is much variability in holding times. Again, it is apparent that the laboratories are either the source of guidance for sample handling or are relied upon for provision of proper containers and handling.
- Agriculture end uses are most common for both Class A and B materials with landscaping second among Class A producers.
- Samplers are interested in gaining a better understanding of the real purpose for the sampling protocol/activity to be performed and are somewhat concerned about the health risks associated with sampling. Despite this expressed position, few feel they require much training based on responses to questions about training and supervision.
- QA/QC programs exist for about half of the utilities that responded. However, not many of those programs include anything specific to microbial sampling and analysis. Thus, if guidance is desired and made available, it comes from preexisting documents made available to the utility.
- There appears to be a small correlation between end use, class and presence of a QA/QC plan.
- The responses indicate that there may be even greater variability in sampling practices for cake materials than for liquid. There were limited responses with details on compost sampling, yet when such information was available, there appeared to be no consistency in depth of sampling or frequency.
- The magic number "7" that is garnered from permit or regulatory guidance documents has influenced protocols put into place based on interpretation of the survey responses.

Although many utility managers and operators appear to understand the importance of sampling for microbial analysis, it is apparent that consistency in training and provision of sampling protocols among utilities could be improved substantially. Sampling staff, if different from the supervisor, manager, or laboratory specialist, may not truly understand the importance of microbial analysis and proper sampling and sample handling procedures. Furthermore, the guidance and requirements that States provide seem to be influencing the level of training and certification pursued.

To develop and implement improved sampling protocols, simplifying technique and training approaches without compromising effectiveness should be considered. Improving the communications among and between professionals within the industry, as well as creating more consistent guidance and training tools, could also help to improve implementation practices and accountability. As part of this project's communications plan, effective means of communicating the importance of appropriate biosolids sampling and handling to safeguard the individuals responsible for sample collection as well as the integrity of the samples were explored. In addition, training materials were developed as was reasonable within budget constraints.

2.3 WEFTEC Workshop

Compared to process design, monitoring and laboratory analysis protocols, the details of sampling and handling of sewage sludge and biosolids has been mostly overlooked in regulatory guidance. A thorough review of practices across the country with respect to sampling and handling has not been recently completed. Therefore, to evaluate the needs and study approaches to investigate appropriate sampling methods, a one-day workshop was convened on October 30th at WEFTEC05 in Washington, D.C., in conjunction with WERF project 04-HHE-7. In addition to the project team, the seven member project subcommittee (PSC), and two WERF project managers, 17 individuals representing academia, wastewater utilities, federal and state government agencies, and commercial laboratories throughout the United States, Canada, and Australia participated. Collectively, the meeting participants represented expertise in applied microbiology, wastewater treatment, environmental pollution, sample analysis, and regulatory compliance. The goal of this workshop was to define the state-of-the-art of biosolids sampling, what is being done well, what needs improvement, and possible strategies for obtaining truly representative samples (demonstrating confidence in microbial analysis results) so that the experimental phases of this project could be optimally designed.

The workshop began with a series of presentations to establish a similar vocabulary among the participants and to spur thoughtful discussion from participants. The workshop opened with a welcome and introduction by the WERF project manager, Bonnie Bailey, Bill Cairns, PSC Chair, and Jim Smith, U.S. EPA.

Rick Stevens of the EPA and Greg Kester of the Wisconsin Department of Natural Resources spoke about the existing regulatory framework for biosolids sampling, including gaps, limitations and challenges. They presented the regulatory perspective of when, where and how to sample to meet public health protection goals. Their presentations included why sampling is important, what is being done well, where there is room for improvement, and what are the challenges to implement and enforce those improvements. Training “the guy with the shovel” to adopt the desired mindset regarding proper sampling is in order here.

Mark Meckes of the EPA presented materials prepared by Robin Oshiro and himself on existing sampling and handling methods from the perspective of using those samples in EPA-approved microbial enumeration methods. They included perspectives on conducting sampling and sampling training that deal with each of the three matrices of interest (liquid, cake, compost) from the sampler's perspective: what is being done well, where there is room for improvement, what are the challenges to being able to implement new sampling methods, and what are the challenges with their particular biosolids matrix.

Mike Jennings of NEIWPCC, who is working with EPA Region I on guidance, spoke on the status of this work. The resulting NEIWPCC guidance document has been discussed elsewhere in this report.

In the afternoon, the principal investigators, Victor D'Amato and Sharon C. Long, presented the status of literature review and utility survey tasks, respectively. They solicited feedback on the usefulness of this information.

Following the presentations, workshop participants split into breakout groups to discuss the major questions and sampling procedures for each major biosolids matrix: liquid, cake, compost and miscellaneous/other. Participants discussed ideas for future research directions and questions for the experimental phase of WERF project 04-HHE-7. The information gathering tasks helped formulate the specific research questions targeted in Phase II of this project (see Chapter 3.0).

2.4 Summary of Phase I Information Gathering

Incorporation of the information gathered in the literature review task, the utility survey, and the presentations and discussions from the WEFTEC05 workshop provided the foundation for the experimental approach for WERF project 04-HHE-7. Based on this information and the brief sampling guidance in Section 8.0: Sampling, Handling and Storage, of proposed EPA methods 1680, 1681 and 1682; a number of general sample collection and sample handling issues and questions emerged.

Sampling topics:

- collection of liquid samples from tanks
- collection of cake samples from dewatering units
- collection of cake samples from drying piles
- collection of compost samples from piles
- sizes and types of sample containers that are optimal
- types of sampling devices

Handling questions:

- Does holding time (zero, six, 24 hours) statistically affect bacterial (fecal coliform and *Salmonella*) and virus (coliphages) enumerations?
- Can pre-sterilized and/or field sterilized containers statistically affect bacterial (fecal coliform and *Salmonella*) and virus (coliphages) enumerations?
- What size and geometry of sample containers optimize cooling of samples?
- Does storage in coolers with blue ice (cooled to -20°C) or wet ice make a difference in sample temperature?

Broad questions pertaining to sampling and handling for obtaining representative, statistically robust measurements:

- What size do discrete samples need to be?
- How many subsamples to be individually analyzed or used to produce composite samples are needed?
- What is the most appropriate grid methodology for sampling over area?
- What are the most appropriate materials and geometry of scoops and samplers?

CHAPTER 3.0

EXPERIMENTAL WORK

The second phase of project work involved conducting controlled sampling, sample handling, and microbial analysis for fecal coliforms and *Salmonella* spp. U.S. EPA methods 1680 and 1682 for biosolids products were used at four (4) facilities producing different types of sludge and biosolids products.

Detection of statistically significant changes in microbial numbers based on sample collection and handling over natural heterogeneity in sludge and biosolids materials was a serious concern discussed at the WEFTEC05 workshop. It was decided to limit the microbial analyses to fecal coliforms for most experiments because the numbers of replicates required to ensure adequate statistical power would be logistically prohibitive for *Salmonella* analysis. The U.S. EPA-approved methods for viruses and helminth ova would have been prohibitively slow and expensive for obtaining numbers of replicate analyses for this project. In addition, the bacterial indicators tend to be present in both untreated sludge and treated biosolids materials in higher numbers than viruses and helminth ova. Several orders of magnitude changes in bacterial numbers through a treatment train are more readily detected when focusing on the bacterial indicators. For selected experiments, *Salmonella* and male-specific coliphage analyses were also conducted to provide information about pathogen and virus numbers and their behaviors.

3.1 Study Site Selection

In accordance with the project goals, the biosolids product matrices included liquid, cake and composted materials. Selection of study sites was based on criteria including proximity to the University of Massachusetts or University of Wisconsin (UMass/UW) to meet six hour holding time requirements, and ability to obtain liquid, cake and/or compost products to meet the focus matrices of this project. Table 3-1 summarizes the characteristics of the four participating utilities. In order to provide greater understanding of the sludge/biosolids utilized in the Phase II experiments, the process descriptions of each of the four study sites are provided.

Table 3-1. Summary of Phase II Study Sites.

	Type of product(s)
Facility #1	• Thickened sludge (liquid) • Belt dewatered cake
Facility #2	• Thickened sludge (liquid) • Centrifuge dewatered cake
Facility #3	• Wood chip compost
Facility #4	• Anaerobically digested, gravity belt thickened liquid biosolids

3.1.1 Facility #1

The average daily flow at Facility #1 is approximately 38.5 million gallons per day, and the facility produces over 55,000 tons of solids per year that are sent for incineration. The sludge tested was produced at the plant by blending thickened activated sludge, primary sludge, imported sludge and scum. Polymer is added to the blended material to aid dewatering, prior to being sent to belt filter presses. The resulting cake averages 22% solids. The cake solids are sent to incineration for heat recovery; therefore, processes to achieve Class A or Class B materials are not applied. Figure 3-1 contains a schematic of the process flow at this facility. Liquid samples were collected as the thickened sludge was being conveyed to holding tanks. Cake samples were collected off the conveyor from the belt filter press to storage.

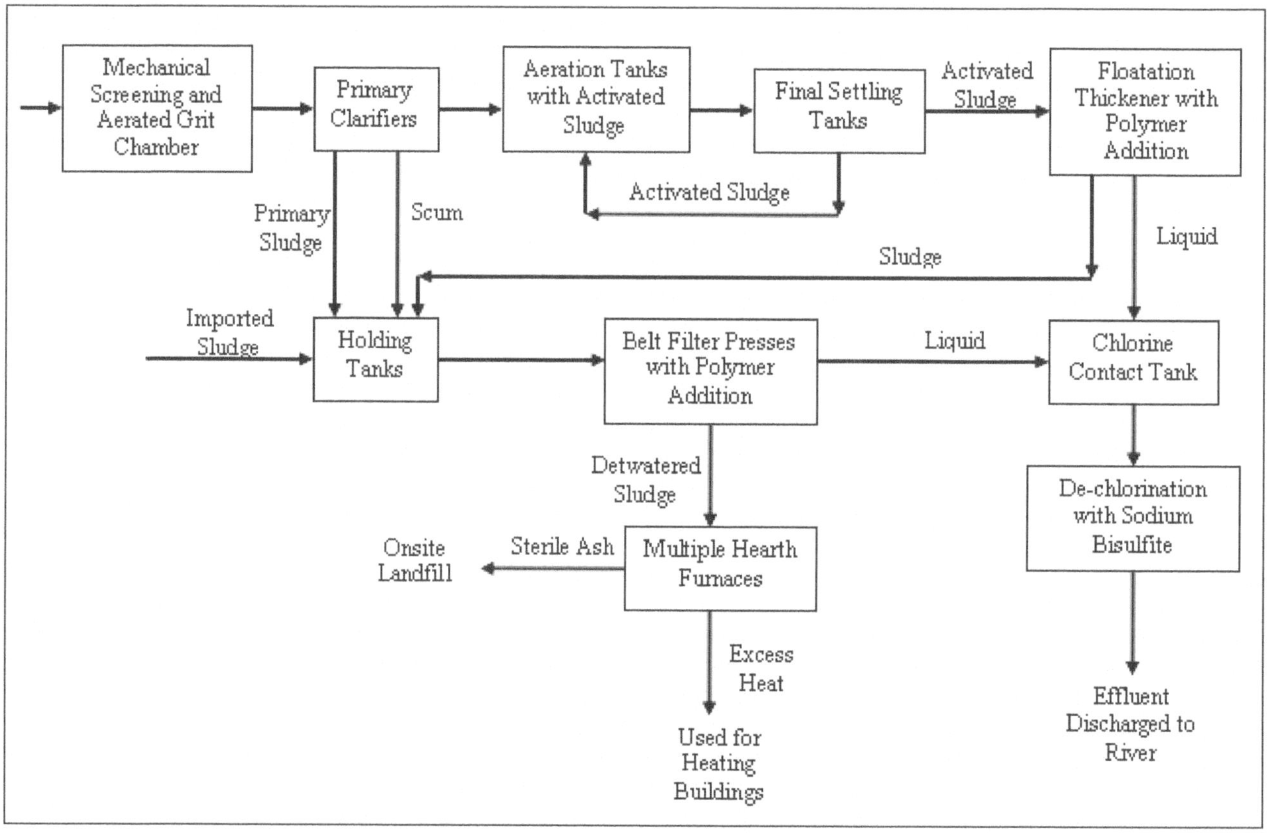

Figure 3-1. Process Flow Schematic of Facility #1.

3.1.2 Facility #2

The average daily flow at Facility #2 is 8.4 million gallons per day, and the plant's process is described as advanced secondary wastewater treatment. A one-stage ammonia-nitrogen removal process is employed. Typical hydraulic flow at this plant begins with wastewater entering the headworks, then into primary settling, followed by aeration/biological treatment, secondary settling, and final disinfection. Blended primary and secondary sludge is gravity thickened and pumped to centrifuges for dewatering. The centrifuge cake, averaging between 20 and 25% solids is then taken to landfill for disposal. The plant processes approximately 1,700 dry metric tons of solids per year. Since the solids are sent for secure disposal, processes to achieve Class A or Class B materials are not applied. Figure 3-2 contains a schematic of the process flow at this facility. Liquid samples were collected from a side spigot off the pipe feeding thickened solids to the centrifuge. Cake samples were collected after centrifugation as the material was being conveyed to storage tanks.

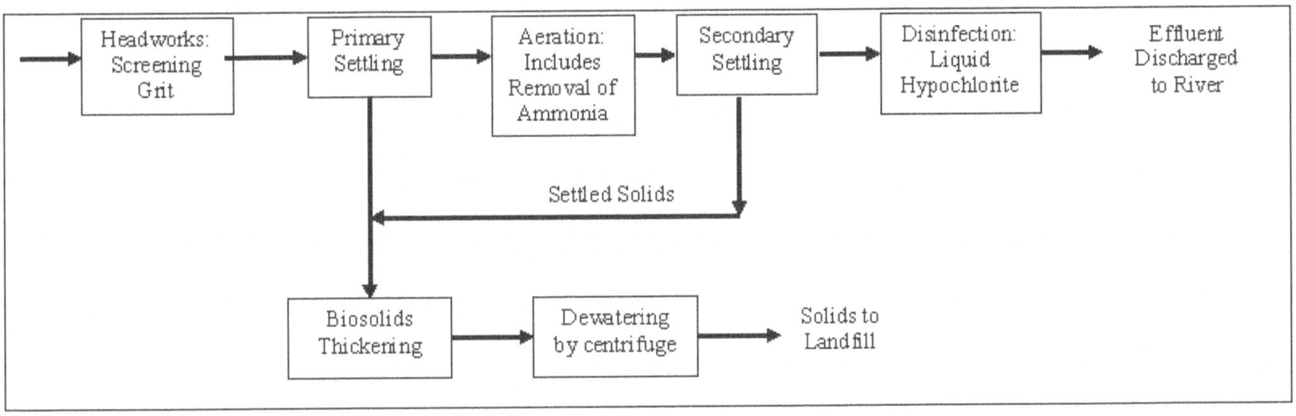

Figure 3-2. Process Flow Schematic of Facility #2.

3.1.3 Facility #3

Facility #3 employs traditional activated sludge secondary treatment with enhanced biological phosphorus removal, treating approximately 1.8 million gallons per day with a design flow of 2.8 million gallons per day. Primary solids are mixed with thickened waste activated sludge (WAS) and anaerobically digested in a two-stage process. The digested solids are pumped to a holding tank. Prior to composting via aerated static piles, the digested solids are dewatered in a rotary drum press and mixed with commercial woodchips. Compost pile temperature is monitored daily. Once pathogen destruction and vector attraction reduction (VAR) time and temperatures are achieved, the compost is screened to recover the larger woodchips. The final biosolids materials are sampled and tested for *Salmonella.* The biosolids are available for community landscaping and municipal uses. Figure 3-3 contains a schematic of the process flow at this facility. Compost samples were collected from piles of finished Class A compost available for garden and landscape use.

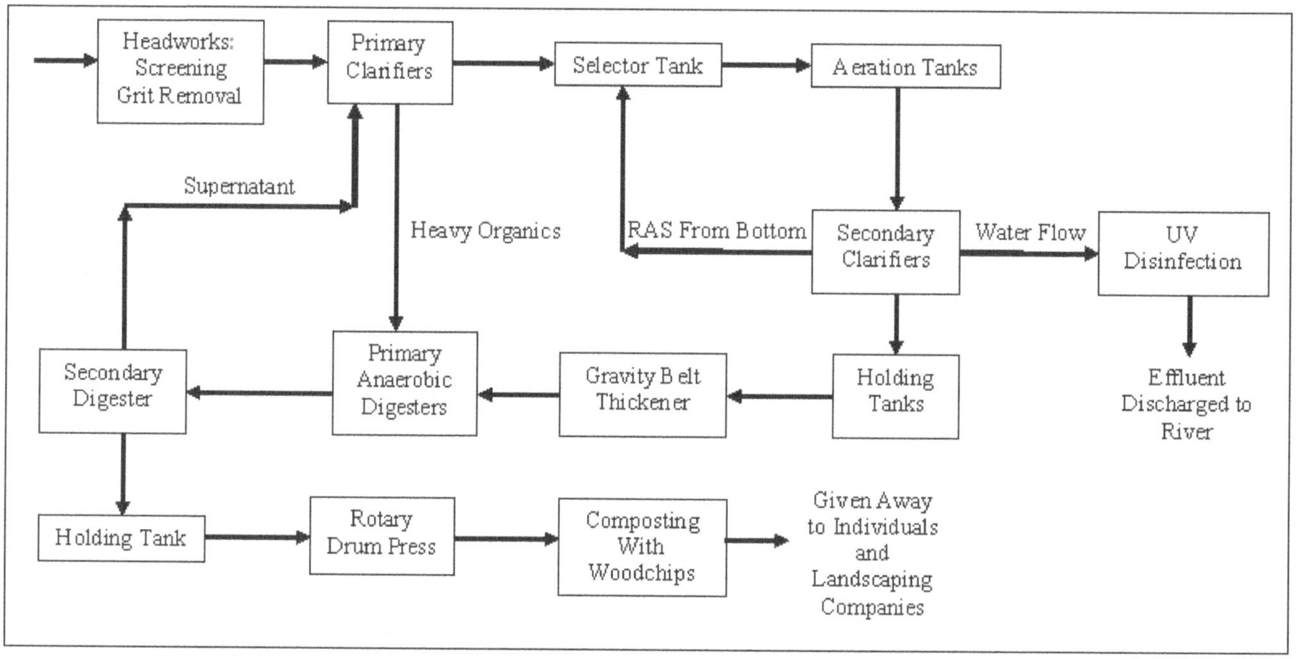

Figure 3-3. Process Flow Schematic of Facility #3.

3.1.4 Facility #4

Facility #4 processes an average daily flow of 42 million gallons per day via conventional activated sludge treatment. As of late summer 2006, mesophilic anaerobic digestion followed by gravity belt thickening was applied to produce a Class B product. At the time of sampling, digested solids were gravity belt thickened and sent to storage tanks. From the storage tanks, the Class B product was transferred to tanker trucks and the biosolids were land applied to agricultural fields. During the time that the facility was providing sampling access, it was undergoing a number of upgrades. The facility was transitioning from mesophilic anaerobic digestion to a phased anaerobic digestion process followed by centrifuge dewatering. Ultimately the facility will produce a Class A product that will be amended with sand and sawdust for texture. Figure 3-4 contains a schematic of the process flow at this facility at the time of sample collection. Liquid samples for analyses were collected from the gravity belt thickener.

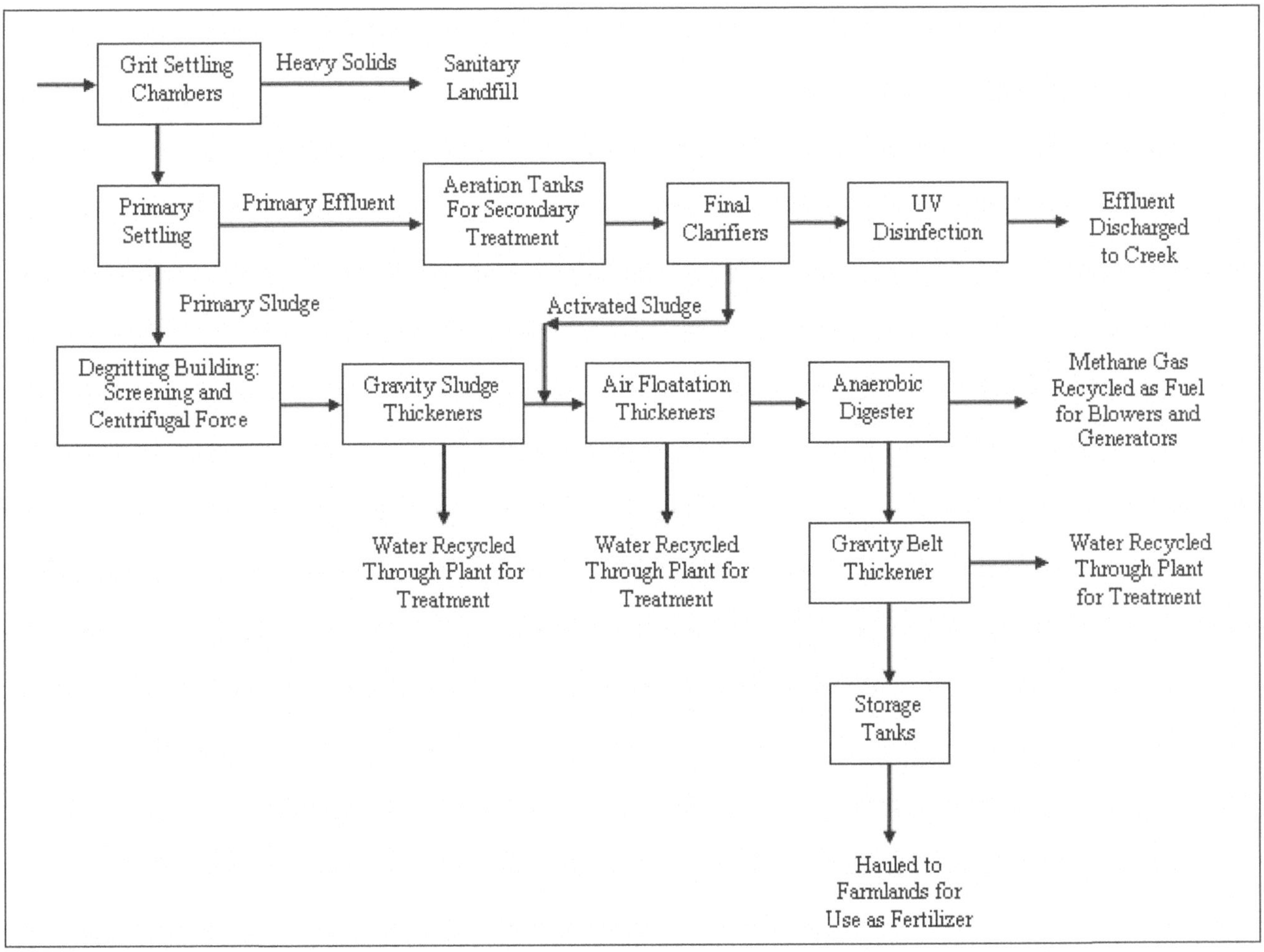

Figure 3-4. Process Flow Schematic of Facility #4.

3.2 Sample Collection and Handling Experimental Approach

The experimental design approach was divided into two categories of questions: sampling-related parameters and handling-related parameters. The information gathered in the literature review, the utility survey, and the presentations and discussions from the WEFTEC workshop provided the foundation for the experimental approach for this project. The Phase II experimental plan also incorporated the information presented at the WEFTEC05 workshop and sampling and handling information provided in EPA methods 1680, 1681, and 1682.

Sampling topics included:

- collection of liquid samples from tanks
- collection of cake samples from dewatering units
- collection of compost samples from piles
- grab versus composite samples
- size and types of sample containers that are optimal.

Handling topics include:

- sample holding time
- sampling cooling/shipping conditions

Based on these topics, the Phase II experimental work focused on a series of research questions involving appropriate sizes of discrete samples, geometry of sample containers, composite sampling procedures, sample holding times and sample cooling regimes. The specific research questions were:

1. How long/how much volume should be purged from pipes before a representative liquid sample can be collected?
2. Does discrete sample size affect sampling outcome?
3. Are composite samples necessary for liquid products stored in tanks?
4. How many increments (subsamples) are required to achieve representative composite samples from conveyors or piles?
5. Does sample holding time (up to 24 hours) affect enumeration outcome?
6. What are the interactions between sample size, sample type and cooling regime?
7. Can field sterilization techniques be used in the collection of aseptic samples as effectively as freshly washed, wrapped and autoclaved equipment?
8. Is sample character affected by the type of sampling device used? Does removing the surface layer of a pile prior to sampling affect results?
9. When multiple laboratories analyze split samples using EPA-approved methods, are the microbial densities comparable?

The experimental approaches used to answer these questions are described below.

3.2.1 Research Question 1

How long/how much volume should be purged from pipes before a representative liquid sample can be collected? Common sense and guidance used in Wisconsin promotes purging of pipes before samples of liquids from tanks should be collected, as presented by Greg Kester (WI DNR) at the WEFTEC05 workshop. The language on how much or how long a pipe should be purged before a sample is collected is vague. Section 8.4 of EPA Methods 1680, 1681 and 1682 state that the sampler should "purge the discharge pipe of old biosolids and warm to the digester temperature by allowing biosolids to flow through the pipe into a container or waste collection device." Depending on plant design, the ability to easily waste the appropriate amount of liquid could be problematic. For example, if the pipe diameter is large, the absence of a drain would require appropriately sized receptacles for capturing and subsequently transporting the wasted material.

The experimental approach was to collect samples from two facilities. The pipe diameters and flow velocities (Q) of the pipes were measured. The volumes present in the side tape pipes were calculated by using the inside diameter of the pipe and length. One pipe volume was determined to be equal to $\pi r^2 l$, where r is the radius of the inside of the pipe and l is the length of the side tap. Samples were collected from the first flush and after various measured wasted volumes. Samples were collected in 1L, wide-mouth, sterile, HDPE bottles. As each sample was collected, the bottle was placed into a cooler containing ice. Once all samples had been collected, they were taken to the onsite laboratory and analyzed for fecal coliform densities using EPA method 1680. Differences among waste and regulatory compliant samples were compared statistically.

3.2.2 Research Question 2

Does discrete sample size affect sampling outcome? EPA Methods 1680, 1681 and 1682 nominally require about 50g of sample for a single analysis (30g for homogenation, 20g for dry weight analysis). Sampling guidance in Section 8 of those methods indicate the use of 1L sterile bags (not of the Whirl-Pak® variety) for liquid samples. No specific container type – with the exception of specifying glass or plastic - is identified for sample collection from conveyor, bin, drying bed or similar samples. The utility survey for this project indicated that utilities are using of a variety of sample containers: 120 mL, 8 ounces, 1 quart to 5 gallons; with many using 1-gallon ziplock-style bags.

To analyze the importance of discrete sample size, samples of different sizes (1L, 2L, or 4L) were collected and replicate subsamples of well mixed samples were analyzed for fecal coliforms. The resulting dataset was examined using hypothesis testing of whether the sample enumerations represent the same population mean. Comparisons of the mean values and associated coefficients of variation from each approach were analyzed to determine which size sample best captures product heterogeneity.

3.2.3 Research Question 3

Are composite samples necessary for liquid products stored in tanks? Discussions during the WEFTEC05 workshop and information from the literature review indicate that single liquid samples are typically collected for analysis. This approach assumes proper mixing of tanks prior to sample collection. The utility survey indicated that some facilities analyze up to seven individual samples per batch. This is derived from the guidance in the 503 Regulations, which requires the analysis of seven samples within a 14-day period for Class B biosolids. However, there was no indication from the variety of information sources evaluated that discrete liquid samples were regularly composited. Fecal coliform enumerations of multiple liquid samples collected on the same day were thus analyzed in order to determine variability among these discrete samples.

3.2.4 Research Question 4

How many increments (subsamples) are required to achieve representative composite samples from conveyors or piles? Based on the literature review of a number of different disciplines that involve sampling, the consensus is that composite samples are needed to obtain representative samples from conveyors and piles. Based on information from the literature, survey results and workshop discussions, composites may typically be produced from four to 50 discrete samples. From the biosolids-related literature, 42 is the upper extent of subsamples cited. Solid and hazardous waste literature indicated that in the Netherlands, 50 increment (grab) samples were used to prepare composites for sampling contaminated soil piles (Lame et al. 2004). From the survey, it appears that the numbers 4 and 7 have some implied significance.

In sampling centrifuge dewatered cake, Lambkin et al. (2004) compared four sample collection methods. Those methods included: 1) collecting nine samples from the centrifuge output, 2) collecting 14 samples from a cake heap using a trowel at 1 to 25 cm, 3) collecting seven samples at 25 to 50 cm from the side of the heap using a corer, and 4) using an excavator to cone and quarter the entire pile, then manually collecting four samples. All samples were composited for each collection method separately. The coefficient of variation within each sampling method was less than 6% for metals, the parameters tested. There was no significant difference at the 95% confidence level in the mean metal content.

For compost piles, samples must be representative of the heterogeneity of the pile. Greater heterogeneity is expected relative to other biosolids matrices due to differences in composting activity because of pile geometry, localized variations in nutrients and moisture content, and other factors. The WEFTEC05 workshop revealed guidance that suggests compositing pile samples by combining discrete samples, mixing by cone and quartering, and continuing the process until the sample is down to a useable size (U.S. EPA, 1993a). The utility survey indicates that many facilities collect samples from piles by digging down into the pile, sometimes 6 inches, sometimes an unspecified depth, with a non-sterile garden trowel or shovel before collecting the sample. It was reported that one to eight samples per batch are collected and analyzed, with the most frequently reported number being seven. As indicated in the literature, the heterogeneity of dewatered cake can be represented by as few as four increments combined in a composite.

Experiments were conducted where seven individual samples were collected using a random grid pattern on a compost pile. Each bottle was placed in a cooler containing ice as soon as collected. The individual samples and a composite of the seven samples were enumerated for fecal coliforms. Comparisons of the mean values and coefficients of variation were used to determine whether seven discrete samples are required to capture sample heterogeneity.

3.2.5 Research Question 5

Do sample holding times affect enumeration outcomes if conducted within 24 hours? EPA-recommended methods for conducting microbial analyses of biosolids advise immediate analysis, usually within 2 hours of collection. In other guidance, it is stated that if enumeration media cannot be inoculated immediately, it should be completed within 6 hours.

In the utility survey, a range of holding times was reported, including 0, 2 to 3, 8, 12, and 16 hours; less than 2 days, and up to 1 week. Approximately 92% of the respondents shipped collected samples within 24 hours, with the remaining 8% shipping within 7 days. Some confusion might arise because holding times for metals analyses are considerably longer than those suggested for microorganisms.

Strict, short holding time requirements essentially necessitate that all facilities for which land application of their biosolids product is a viable option must have an onsite laboratory. This may not be a significant issue if only fecal coliforms are to be analyzed. However, analysis for *Salmonella,* infectious viruses and helminth ova require staff with specialized and less widely available training, in addition to various expensive equipment. For example, to reach the completed stage of *Salmonella* testing, multiple different incubation temperatures are required, and onsite analytical capabilities for all four microbial targets may not be feasible for wastewater treatment plant laboratories, especially smaller facilities. If storage in a cooler for 24 hours does not statistically affect enumeration results, then overnight shipping of samples to a contract laboratory would be permissible. Part 503 and EPA guidance allows a 24 hour holding time for only certain types of biosolids that will be analyzed using Methods 1680 and 1681 (U.S. EPA, 2006c).

For this project, one-liter samples of sludge and biosolids were collected and placed on ice. Samples were transported to the onsite laboratories of the study facilities, and duplicate enumeration series for fecal coliforms using EPA method 1680 were prepared within the 2-hour limit. All tubes were placed in a portable 35°C incubator for transport. Upon return to the project team's laboratory, a second duplicate enumeration series for fecal coliforms using EPA method 1680 was prepared at approximately 6 hours after sampling. Both 0 to 2 hour and 6-hour MPN sets were then incubated in the 35°C laboratory incubator. Remaining sample was stored overnight in a cooler and the cooler temperature was recorded. A third and final duplicate enumeration series for fecal coliforms, using EPA method 1680, from subsets of the sample was inoculated within 24 hours. This experiment was repeated for samples from multiple facilities. Statistical analyses were conducted to determine the significance of holding time on enumeration values.

3.2.6 Research Question 6

What are the interactions of sample size, sample type and cooling regime? There was some variation in guidance regarding the length of time samples are allowed to be stored or in transit to a laboratory before analysis, and how to cool the sample during storage and shipping. The *Federal Register* [72 *Fed. Reg.* 57 (26 March 2007)] indicates that samples intended for fecal coliform and *Salmonella* analyses are to be cooled to <10°C with a maximum holding time of 6 hours, although it is preferable if sample analyses are initiated within 2 hours of collection. However, holding times may be extended to 24 hours for biosolids analyzed using Methods 1680 or 1681 for Class A compost or digested (anaerobic or aerobic) Class B products.

It should be noted that these holding times are either less than or the same as those in the *White House Document*, but with slightly different temperatures for the Part 136 methods. Specifically, if using Methods 1680 or 1681 for fecal coliforms, and assaying Class A composted, Class B aerobically digested or Class B anaerobically digested biosolids, the holding time is 24 hours. All other combinations for the three EPA methods (1680, 1681, and 1682) are 6 hours. Also, the holding temp is <10ºC, not frozen for all three EPA methods for all sample types.

These handling requirements are intended to prevent changes in the physiological state of the microbial targets during storage and transport. As the analytical methods for microbial targets in biosolids all rely on culture-based methods, or a positive growth end point, changes in physiological state can result in changes in enumeration values. If injury continues during storage and transport, lower than actual organism densities are likely to be measured. If conditions allow for regrowth, higher than actual organism densities may be measured. Either way, the resulting values would not be representative of the sampled biosolids.

To investigate the affects of cooler packing and storage, a number of experiments were conducted to assess how rapidly the core of the sludge and biosolids samples fell below 10ºC. In combination with experiments conducted to address research questions 2 and 5, the interactions between sample size, sample type and cooling regime were examined.

3.2.7 Research Question 7

Can field sterilization techniques allow the collection of aseptic samples as effectively as freshly washed, wrapped and autoclaved equipment? To put this question into context, consider the use of laboratory blenders for sample homogenation. Due presumably to budget constraints (and possibly lack of awareness), many utility laboratories own blenders that can be purchased at retail stores for about $50-100. The gaskets on these retail blenders generally do not allow the blender container to be autoclaved. Stainless steel blender containers that can be autoclaved are available from scientific suppliers, such as Fisher Scientific and VWR, for approximately $250. Should a utility need to analyze several biosolids samples in a single day and utilize autoclaved blender containers, they may need to invest in several blender containers, in addition to at least one blender base, at additional cost.

In previous work by the principal investigator, it was demonstrated that triple washing blender containers with three 300 mL aliquots of 50% (v/v) ethanol/sterilized water, followed by three 300 mL aliquots of sterile laboratory water, resulted in no growth when nutrient broth was added and incubated overnight at 35°C.

In this project, experiments were conducted with retail blenders using the above-described sterilization scheme and tested by incubating with lauryl tryptose broth, relevant to fecal coliform analysis by EPA method 1680. In addition, biosolids samples were enumerated using an autoclave sterilized stainless steel blender. The blender was cleaned and sterilized using 50%(v/v) ethanol/water and sterilized water rinses, and the same biosolids were enumerated again.

These experiments were conducted to potentially demonstrate that such a scheme could be used to "field" sterilize other sampling devices such as trowels and shovels. Some slight error associated with sample moisture content (and thus dry weight corrections) could be introduced if the sampling device was used before dry. However, since the MPN enumeration techniques provide order-of-magnitude estimates of microbial density, this error does not seem to be significant.

3.2.8 Research Question 8

Is sample character affected by the type of sampling device used? EPA specifies that sample utensils or sampling devices should be made of stainless steel or plastic, but never galvanized or zinc-coated materials. EPA guidance also notes that the geometry of scoops and samplers can affect the nature of the sample obtained. For instance, information gathered in the literature review and discussed at the WEFTEC05 workshop indicated that the use of a curved scoop can bias a sample, such as compost, toward exclusion of large pieces. Guidance for producing homogenized samples in EPA Methods 1680, 1681, and 1682 addresses exclusion of large particles, such as branches from the subsample.

Owing to budget limitations for this project, the testing of large sampling devices, typically used for piles and drying beds would have been prohibitively expensive (the cost of purchasing numerous different devices). Therefore, to begin to answer these questions, fecal coliform enumerations of large and small compost particles were conducted. In addition, enumerations of samples from the 6-inch surface layer of a pile were compared with those taken from piles after removing 6 inches off the surface. For Phase III of this project, utilities that employ large-scale devices were included and different sample collection schemes were tested at those facilities.

3.2.9 Research Question 9

When multiple laboratories analyze split samples using EPA-approved methods, are the microbial densities comparable? Recently approved EPA Methods 1680, 1681, and 1682 contain aspects of the methods that are new to many utility laboratories. A number of experiments were conducted between this project team and the laboratory staff at Facility #4 to determine whether analyzing split and duplicate samples using these methods yield acceptable variability in results.

3.2.10 Statistical Analyses

For the parameters being tested, the most-probable-number (MPN) values were imported into SAS. For a number of research questions (pipe purge, holding time, sample size, and particle size) a regression model for the $\log_{10}$(MPN) value was derived. By evaluating the p-values for the models against the mean squared error (MSE), the significance of the factors tested could be evaluated. For the compost experiment where shallow and deep samples were compared, a paired *t*-test was used. In addition, the data were entered into SAS Mixed Procedure grouping variables such as early pipe purge vs. later pipe purge, sample holding time, or sample volume. The estimated mean values were compared using non-parametric test output (which does not assume normality in the data) to determine whether the factor tested was significant. Results of corresponding statistical analyses are presented for each experiment, as appropriate, below.

3.3 Experimental Results

3.3.1 Pipe Purging for Liquid Samples

These experiments focused on purging of pipes prior to collecting liquid samples. As discussed above, current liquid sampling guidance regarding how much liquid or how long a side tap should be purged before collecting a sample is vague. This experiment was first conducted at Facility #1. The belt-filter presses at Facility #1 are fed with thickened sludge and the liquid samples are collected from a side valve. The diameter of the feed pipe is approximately 5 inches and one pipe volume was approximately 750mL. A bucket marked in liter increments was placed under the pipe, and 1L samples were collected by placing the sample container in the flow after approximately 0, 2, 3, 4.25, 7, 8.25, 10 and 11.25 L of wasted liquid had collected in the bucket. The material collected in the bucket was not used. Approximately 20 minutes elapsed during sample collection. The 1L sample bottles were immediately placed in coolers, and processed in the Facility #1 laboratory within 2 hours after collection.

The results of the enumerations are presented in Figure 3-5. Values ranged from 1.1×10^7 to 7.5×10^7 MPN/g dry wt, less than one order of magnitude. The geometric mean for the first four samples, less than 10 pipe volumes, was 3.0×10^7 MPN/g dry wt. The geometric mean for the last three volumes, more than 10 pipe volumes, was 2.3×10^7 MPN/g dry wt. The geometric mean for all seven samples was 2.7×10^7 MPN/g dry wt. Hypothesis testing indicated that the geometric means of the two sample sets (<10 L vs. >10 L passed) were equal at $\alpha=0.5$, or 95% confidence. Note that four samples were collected after 10 pipe volumes had passed; however, owing to a laboratory accident, only data from three samples could reliably be reported.

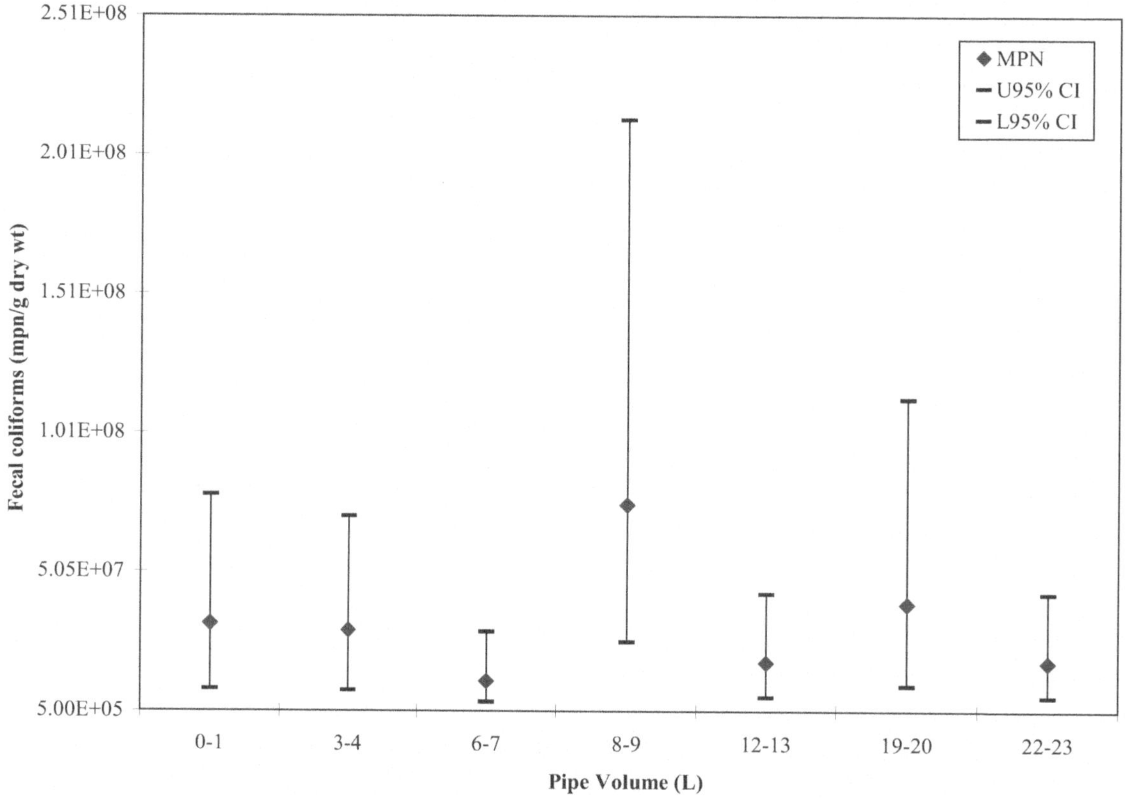

Figure 3-5. Results of Facility #1 Pipe Purge Experiment

This experiment was repeated at Facility #2. Samples were collected from a side valve from the pipe that normally feeds the centrifuges. The wasted pipe volumes varied from the previous experiment because the feed pipe was of a smaller diameter than at Facility #1. Again, the 1L sample bottles were immediately placed in coolers and processed in the Facility #2 laboratory within two hours of collection. The enumeration results are presented in Figure 3-6. All samples contained on the order of 10^6 MPN/g dry weight fecal coliform. The geometric mean for the first four samples, less than 10 pipe volumes, was 1.6×10^6 MPN/g dry wt. The geometric mean for the last four samples, greater than 10 pipe volumes, was 2.2×10^6 MPN/g dry wt. The geometric mean for all eight samples was 1.9×10^6 MPN/g dry wt. Hypothesis testing indicated that the geometric means of the two sample sets were equal at α=0.5, or 95% confidence. One apparent outlier sample was collected after 16 to 17 liters had passed out of the pipe.

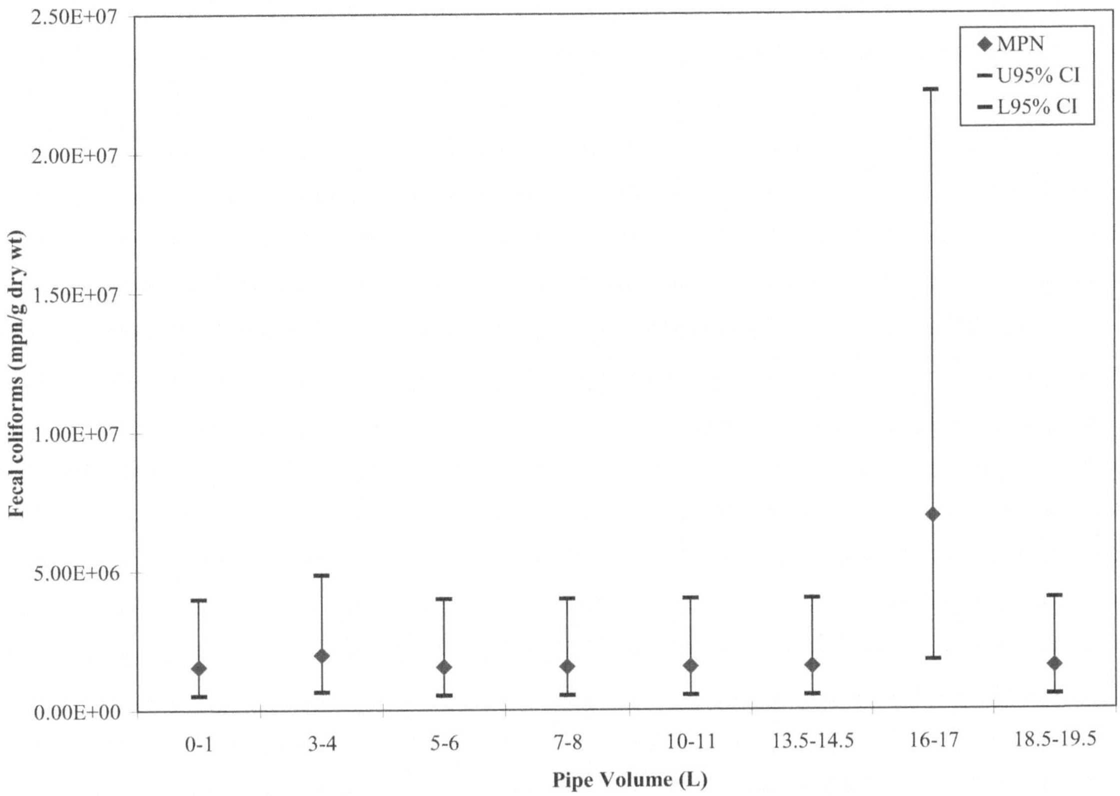

Figure 3-6. Results of Facility #2 Pipe Purge Experiment.

The results of this experiment suggest that multiple samples of liquid sludge or biosolids be collected and analyzed if time and budgets allow. In this manner, greater insight into product batch variability, such as seen in the 16 to 17L sample, will more likely be captured. A greater understanding of the variability in sludge or biosolids microbial content will help to assure confidence in product safety.

It is important to note that the inclination of both operators that assisted with liquid sampling was to collect the sample immediately upon opening the valve. Neither location could easily accommodate liquid wasting, as no floor drain was available within reasonable proximity to the pipes. This observation indicates the need for careful education of the sample collectors and underscores the need for engineers/designers to be cognizant of sampling needs. Provisions are needed to ensure that wasting of appropriate amounts of liquid sludge or biosolids can be achieved nearly effortlessly.

All pipe purge experimental results are summarized again in Table 3-2. A qualitative evaluation of the data indicates that proper purging of side stream pipes was not a significant factor in these experiments. The number of positive tubes in the 5x5 sets of dilutions varied between samples typically by no more than two or three tubes. Multiple regression analysis using SAS to model whether certain test variables had a significant affect on the MPN/g dry wt outcome was performed. The variable of significance for the pipe purge experiments was whether or not the samples were valid EPA samples, collected after 10 pipe volumes were purged. This analysis was conducted for the Facility #1 and Facility #2 datasets combined, and for each separately.

The p-value for the combined dataset was 0.4120 which would indicate that the MPN values for EPA-valid and EPA-invalid samples were not statistically different. However, the mean squared error (MSE) and square root of MSE are 0.158 and 0.397, respectively. Since the error in the MSE is larger than the estimated variance (s^2) of the MPN values of 0.0544 by only a factor of 2.9, the strength of this conclusion is relatively weak. Ideally, s^2 should be much smaller than MSE to lend strength to the conclusions drawn from this type of statistical analysis. Similar results were noted when data from each treatment plant were analyzed separately.

In reviewing the SAS output, the estimated mean values for EPA-valid samples were always the largest among the two treatment groups considered. Therefore, there might be a small effect on sample enumeration if little or no material is wasted from the pipe before sampling. Because of the large confidence intervals generated by the MPN method, this conclusion could not be supported statistically. It is important to note that facility (#1 versus #2) was demonstrated to be a significant factor in determining final MPN value. This conclusion lends credibility to the statistical analysis model, since the quality of liquid sludge at each facility would be different and one would expect their MPN values to be different.

Table 3-2. Summary of Pipe Purge Experiment Data.

Facility #1 (volume passed)	MPN/g dry wt	95% CI
0-1L	3.2×10^7	$8.3 \times 10^6 - 7.8 \times 10^7$
3-4L	2.9×10^7	$7.9 \times 10^6 - 7.0 \times 10^7$
6-7L	1.1×10^7	$7.8 \times 10^6 - 2.9 \times 10^7$
8-9L	7.5×10^7	$2.5 \times 10^7 - 2.1 \times 10^8$
12-13L	1.8×10^7	$5.6 \times 10^6 - 4.3 \times 10^7$
19-20L	3.9×10^7	$9.7 \times 10^6 - 1.1 \times 10^8$
22-23L	1.8×10^7	$5.6 \times 10^6 - 4.3 \times 10^7$
Facility #2 (volume passed)		
0-1L	1.5×10^6	$5.2 \times 10^5 - 4.0 \times 10^6$
3-4L	2.0×10^6	$6.5 \times 10^5 - 4.9 \times 10^6$
5-6L	1.5×10^6	$5.2 \times 10^5 - 4.0 \times 10^6$
7-8L	1.5×10^6	$5.2 \times 10^5 - 4.0 \times 10^6$
10-11L	1.5×10^6	$5.2 \times 10^5 - 4.0 \times 10^6$
13.5-14.5L	1.5×10^6	$5.2 \times 10^5 - 4.0 \times 10^6$
16-17L	6.9×10^6	$1.8 \times 10^6 - 2.2 \times 10^7$
18.5-19.5L	1.5×10^6	$5.2 \times 10^5 - 4.0 \times 10^6$

3.3.2 Effects of Discrete Sample Size

It can be hypothesized that collecting larger, well-mixed samples would incrementally yield results more representative of the biosolids batch. At Facility #1, discrete samples of 1L, 2L and 4L of belt filter press dewatered cake were collected directly from the sample belts and enumerated in duplicate using Method 1680. Figure 3-7 presents the fecal coliform results. Variability between replicate (R) enumerations of the same sample was greater than variability across sample sizes. This result suggests that 1L samples are adequate for microbial analysis. However, the variability for the 4L samples appears to be the smallest when sample size replicates are compared to only its paired sample.

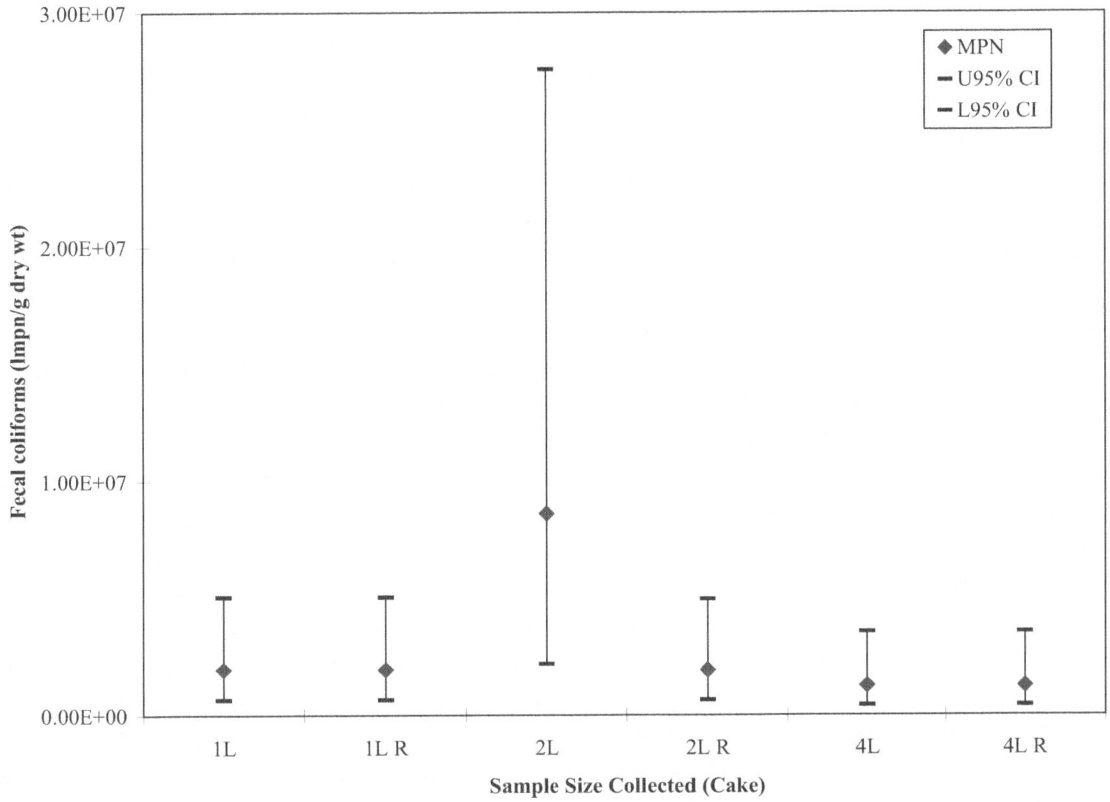

Figure 3-7. Effect of Discrete Sample Size from Facility #1.

At Facility #4, discrete samples of 1L, 2L and 4L of gravity thickened liquid were collected and enumerated in triplicate by EPA Method 1680. All sample bottles were placed in coolers and surrounded by gel packs on all four sides. The results of Facility #4 sample size enumerations are presented in Figure 3-8. Once again, variability among primary, duplicate (D), and triplicate (T) enumerations of the same sample was greater than variability across sample sizes. This result suggests that 1L samples are adequate for microbial analysis.

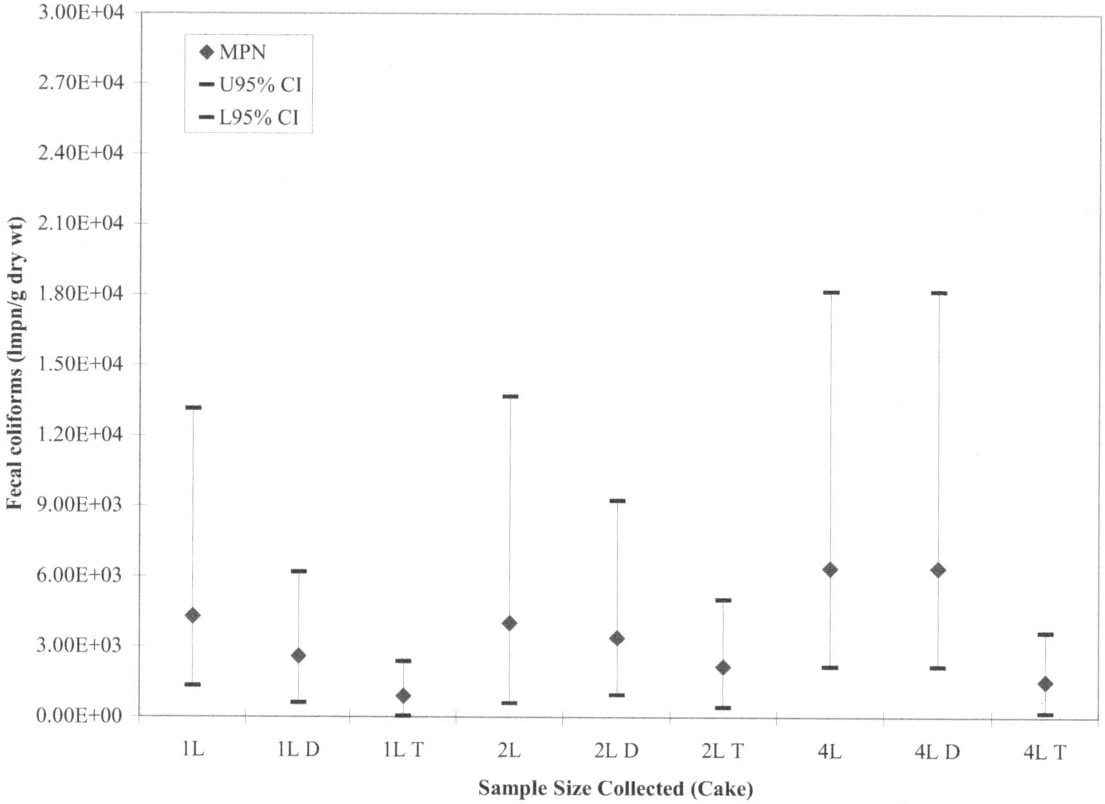

Figure 3-8. Effect of Discrete Sample Size from Facility #4.

The results from the sample size experiments are summarized in Table 3-3. Upon qualitative observation of the data, the series of positive tubes in each 5x5 set of dilutions varied by up to four tubes. The final numbers were all within a single order of magnitude of each other, and a number of the replicates yielded identical results. Multiple regression models were evaluated in SAS with sample size as a variable of interest. "Facility" was demonstrated to be significant in affecting final MPN values. When considering facilities separately, the MSE for each model were approximately equal to the variance, s^2. Therefore, while the p-values from each model indicated that sample size was not statistically significant, the statistical strength of that conclusion was weak. It is interesting to note that staff from Facility #4 conducted sampling and analysis of the gravity belt thickened liquid, collecting a sample in a 2L container three days before the samples collected and analyzed in this study, and reported $3.7\text{x}10^3$ MPN/g dry weight solids, directly comparable to the numbers reported in Table 3-3.

Table 3-3. Summary of Sample Volume Experiment Data.

Facility #1 cake	MPN/g dry wt	95% CI
1L	$2.0x10^6$	$6.6x10^5 – 5.1x10^6$
	$2.0x10^6$	$6.6x10^5 – 5.1x10^6$
Geometric mean	$2.0x10^6$	
2L	$8.6x10^6$	$2.2x10^6 – 2.8x10^7$
	$1.9x10^6$	$6.5x10^5 – 5.0x10^6$
Geometric mean	$4.1x10^6$	
4L	$1.3x10^6$	$4.2x10^5 – 3.6x10^6$
	$1.3x10^6$	$4.2x10^5 – 3.6x10^6$
Geometric mean	$1.3x10^6$	
Facility #4 liquid	MPN/g dry wt	95% CI
1L	$4.3x10^3$	$1.3x10^3 – 1.3x10^4$
	$2.6x10^3$	$6.2x10^2 – 6.2x10^3$
	$8.9x10^2$	$6.0x10^1 – 2.4x10^3$
Geometric mean	$2.1x10^3$	
2L	$4.0x10^3$	$6.0x10^2 – 1.4x10^4$
	$3.4x10^3$	$1.0x10^3 – 9.3x10^3$
	$2.2x10^3$	$4.4x10^2 – 5.0x10^3$
Geometric mean	$3.1x10^3$	
4L	$6.4x10^3$	$2.2x10^3 – 1.8x10^4$
	$6.4x10^3$	$2.2x10^3 – 1.8x10^4$
	$1.5x10^3$	$1.9x10^2 – 3.6x10^3$
Geometric mean	$4.0x10^3$	

3.3.3 Composite Sampling for Liquid Biosolids

The 503 Regulations specifically state that composite samples should not be collected or analyzed for microbial analyses; however, they are acceptable for metals and toxic chemical analyses. In order to begin to answer the question regarding the value of producing composites for liquid biosolids for microbial analyses, multiple samples were collected from the gravity belt thickener from Facility #4 over the span of about one hour. Seven samples were analyzed by the project team using EPA Method 1680; one sample (sample #1) was split and analyzed for fecal coliforms in duplicate by the utility. In addition, the project team analyzed three samples using EPA Method 1681 for fecal coliforms. Facility #4 analyzed their sample in duplicate for *Salmonella* using EPA method 1682. The same three samples analyzed via EPA Method 1681 were also measured for *Salmonella* using EPA Method 1682 by the WERF project team. Finally, all seven samples were analyzed by the project team for male-specific coliphages using EPA Method 1602. The results are summarized in Table 3-4.

Table 3-4. Multiple Analyses of Project Team's Liquid Samples.

Sample ID	FC by 1680 (MPN/g dry wt)	95% CI	FC by 1681 (MPN/g dry wt)	95% CI
1	870	58 - 2,330	387	58 - 1,320
2	1,320	116 - 3,210	2,070	425 - 4,840
3	774	58 - 2,090	349	NC
4	1,530	193 - 3,640		
5	349	NC		
6	870	58 - 2,300		
7	2,070	425 - 4,840		
Geometric mean	977		654	
	Salmonella (MPN/4 g dry wt)		F+coliphage (pfu/g dry wt)	
1	5.2	0.9 - 18	ND	
2	ND	0 - 17	65	
3	ND	0 - 17	ND	
4			ND	
5			ND	
6			ND	
7			ND	
Geometric mean	3.5		9.3	

NC - not calculated, no values in MPN table
ND - not detected

The analyses conducted at the Facility #4 laboratory (split of sample 1) yielded 717 and 426 MPN/g dry weight fecal coliforms by EPA Method 1680 in duplicate analyses. For all of the fecal coliform analyses conducted by the WERF project team, the numbers of positive tubes among all the 5x5 tube dilutions varied by up to three tubes with a resulting range of fecal coliform densities of 349 to 2,070 MPN/g dry wt. For *Salmonella,* only a single isolate from three sets of 3x5 dilutions was confirmed. For duplicate analyses conducted by Facility #4 staff (split of sample 1), no *Salmonella* isolates were confirmed. For male-specific or F+coliphages, only one replicate plate contained plaque forming units.

Qualitatively, the fecal coliform values for each sample are contained within the 95% confidence intervals of the other enumerations. This would suggest that composite liquid samples are not necessary, consistent with guidance in the 503 Regulations. However, it was also shown that multiple analyses of discrete liquid samples are useful for detecting low levels of indicators and pathogens, such as was demonstrated for the *Salmonella* and F+coliphage enumerations presented here.

3.3.4 Numbers of Samples and Production of Composite Samples

An experiment involving a number of compost samples was conducted in order to investigate production of composite samples. Samples were collected from the compost storage pile at Facility #3. The compost product is available to the public at no cost at this facility. The dimensions of the pile were measured and divided into a regular 4x4 grid. Seven samples, a number implied as being significant in the 503 Regulations by utility survey respondents, were collected along a Z-pattern through the grid. Each sample was enumerated following EPA Method 1680. In addition, approximately equal volumes of each sample were combined to produce a composite. The resulting composite pile was coned and quartered for approximately three minutes. Then, an appropriate 30g wet weight of sample was enumerated, consistent with EPA Method 1680 guidance. Dry weights were determined on triplicate subsamples. Table 3-5 presents the results of sample enumeration as well as the geometric mean for the seven discrete samples.

Table 3-5. Results of Multiple Compost Sample Analysis.

Sample ID	FC by 1680 (MPN/g dry wt)	95% CI
1	$2.0x10^3$	$6.2x10^2$ - $4.7x10^3$
2	$5.5x10^2$	$1.4x10^2$ - $1.8x10^3$
3	$2.7x10^3$	$7.7x10^2$ - $6.4x10^3$
4	$1.2x10^3$	$4.2x10^2$ - $3.2x10^3$
5	$3.2x10^2$	$8.7x10^1$ - $7.8x10^2$
6	$4.3x10^4$	$1.1x10^4$ - $1.2x10^5$
7	$6.0x10^2$	$1.9x10^2$ - $1.9x10^3$
Geometric mean	$1.6x10^3$	
Composite	$1.2x10^3$	

Excluding sample 6, the samples contained from 324 to 2,710 MPN/g dry wt fecal coliforms, similar to previous experiments demonstrating about an order of magnitude variability, similar to previous experiments. Sample 6 contained 43,000 MPN/g dry wt fecal coliforms. Had only four samples been collected, as discussed in some guidance documents, this pocket of elevated coliforms would have been missed. The composite sample contained 1,230 MPN/g dry wt and the geometric mean of the seven discrete samples was 1,630 MPN/g dry wt. These values were close to the mode value with four samples containing lower and three samples containing higher densities of fecal coliforms. Overall, the data support analysis of multiple discrete samples of materials from piles in order to fully capture the heterogeneity of the material. Conducting analysis on a composite alone would have masked the higher levels enumerated in sample 6 (and sample 3), which are cause for concern when distributing a Class A product.

In conducting *Salmonella* analyses on gravity belt thickened (GBT) liquid samples from Facility #4, enumerating one 30g wet weight aliquot of biosolids resulted in a detection limit of

5.25 MPN/4g dry weight (Table 3-4). Since the detection limit is above the regulatory limit of 3 MPN/4g dry weight, a non-detect could not be conclusively judged to meet Class A requirements. In correspondence with U.S. EPA, it was suggested that larger samples of biosolids would need to be analyzed to achieve detection limits below the regulatory limits for Class A biosolids. If the 30g wet weight/270mL buffer slurry was used to inoculate a 20.0mL, 10.0mL, and 1.0mL series as specified in Method 1682 (U.S. EPA, 2006b), Table 3-6 presents the number of replicate analyses needed on the same liter sample to achieve a detection limit just below 3 MPN/4g dry weight. Conversely, a facility that produces a biosolids product with 3% solids content, a slurry of 90g wet weight/210mL buffer in one set of MPN tubes would yield an acceptable detection limit.

Table 3-6. Replicate Analyses Necessary to Achieve 3 MPN/4 g dry weight *Salmonella* Detection Limits.

Biosolids Solids Content	Number of Replicate Analyses
1%	9
2%	5
3%	3
4%	3
5%	2
6%	2
7%	2
8%	2
9%	1
10%	1

A key finding in this work is that analysis of multiple discrete samples provides greater information regarding the variability in microbial content than does analysis of composite samples. Table 3-7 presents a matrix of sample numbers needed to satisfy a given level of certainty, and an allowable fraction of a biosolids batch that does not meet the desired regulatory classification for microbials (Class A or Class B). Using this table, a utility manager could determine how many discrete samples need to be collected and analyzed to detect an unacceptable batch. For example, if a manager wanted to be 90% certain that their material met Class A, and the chance that there is a hotspot in the batch is 1%, then 230 discrete samples would need to be collected and analyzed. Regular analysis of multiple grab samples from a single biosolids batch will help individual utilities understand the variability and likelihood of microbial hotspots in their product.

Table 3-7. Samples Needed for Various Certainty and Product Consistency.

	Fraction percent of "hotspots"													
Certainty	0.01	0.02	0.03	0.04	0.05	0.06	0.07	0.08	0.09	0.1	0.15	0.2	0.25	0.3
0.5	69	35	23	17	14	12	10	9	8	7	5	4	3	2
0.6	92	46	31	23	18	15	13	11	10	9	6	5	4	3
0.7	120	60	40	30	24	20	17	15	13	12	8	6	5	4
0.8	161	80	53	40	32	27	23	20	18	16	10	8	6	5
0.9	230	114	76	57	45	38	32	28	25	22	15	11	9	7
0.95	299	149	99	74	59	49	42	36	32	29	19	14	11	9
0.99	459	228	152	113	90	75	64	56	49	44	29	21	17	13

Table 3-7 was constructed based on a simple statistical model, as follows. If the probability of a single, discrete, one-liter sample being non-compliant is p, then the probability of getting no unacceptable samples out of n total samples is p^n. Therefore, the probability of collecting at least one bad sample out of n is $1 - p^n$. Choosing a specific value of p (these run along the top of the table as column headings) and then setting this equation equal to a desired certainty (probability of being correct), termed P (which runs along the table's left side as row headings), allows the equation to be solved for n, the main table entries. Equation 1 shows the solution to $1 - p^n = P$.

$$n = \log(1-P) / \log(p) \qquad \text{Equation 1}$$

The sample numbers (n) required were rounded up to the nearest integer.

3.3.5 Effects of Sample Holding Times

EPA-approved methods for biosolids microbial analyses indicate that media should be inoculated immediately upon sample collection, ideally within two hours. However, up to six hours may elapse where delays cannot be avoided. This sample holding time requirement precludes utilities from sending samples to contract labs via overnight shipping for analyses, as this practice typically results in holding times ranging from 18-30 hours depending, on when the sample was collected and packed. For certain biosolids products (Class B aerobically or anaerobically digested biosolids, and Class A compost), samples may be analyzed after 24 hours using Methods 1680 or 1681 only (U.S. EPA, 2006c).

In order to assess the impact of storing samples in coolers for shipping, 2L samples of liquid and belt-filter press cake were collected in sterile HDPE screw capped bottles from Facility #1. One set of LTB MPN tubes was inoculated following EPA Method 1680 at the facility and transported under ambient temperature conditions (approx. 2 hours at 22-25°C, because a portable 35°C incubator was not available) back to the research laboratory for incubation (sample 0). Sample bottles were placed in a cooler containing freezer gel packs and brought back to the research laboratory.

After approximately 6 hours of storage, duplicate series of LTB MPN tubes were inoculated with a second subsample from the 2L bottles. For one set, sample 6S, the tubes were allowed to sit at room temperature (20-23°C) for approximately 2 hours (to mimic a typical transport time for 6-hour tubes) after inoculation. Another set of tubes, sample 6, was incubated immediately after inoculation to assess the impact of transport time on results. A temperature blank in the cooler at 6 hours read 6.7°C. Finally, the coolers were allowed to sit overnight, after which another set of LTB MPN tubes was inoculated with a third subsample from the 2L bottles (sample 12). A temperature blank in the cooler at 24 hours read 8.3°C. For fecal coliform analyses, the Federal Register indicates that coolers must be maintained below 10°C. Therefore, these samples would have been considered compliant.

All series of MPN tubes were incubated, transferred and results recorded consistent with EPA Method 1680. The results of enumerations are summarized in Figure 3-9 for liquid and Figure 3-10 for belt-filter press cake. For the liquid samples, the most probable number of fecal coliforms enumerated appeared to decrease with storage time. However, the 95% confidence limits overlapped significantly. The range of enumerations was $5.6 x 10^6$ to $2.5 x 10^7$ MPN/g dry wt, less than one order of magnitude. For cake, the results were similar. Although levels of fecal coliforms enumerated ranged from $4.8 x 10^5$ to $1.8 x 10^6$ MPN/g dry wt, again less than one order of magnitude, the 95% confidence limits overlapped significantly. While it qualitatively appears that storage for 24 hours can result in enumerating lower FC densities, the difference in numbers is less than the span of the method's 95% confidence limits, suggesting that the results are not statistically significant.

For the liquid sample, sitting for approximately two hours between inoculation and incubation resulted in a higher final MPN (6 vs. 6S). To the contrary, sitting resulted in a lower final MPN measurement for cake samples. The 95% confidence limits included the MPN value for the other treatment (i.e. 6 vs. 6S). Therefore, these differences can be concluded to represent inherent variability in measurement, rather than a significant effect attributable to transporting samples stored at room temperature for two hours.

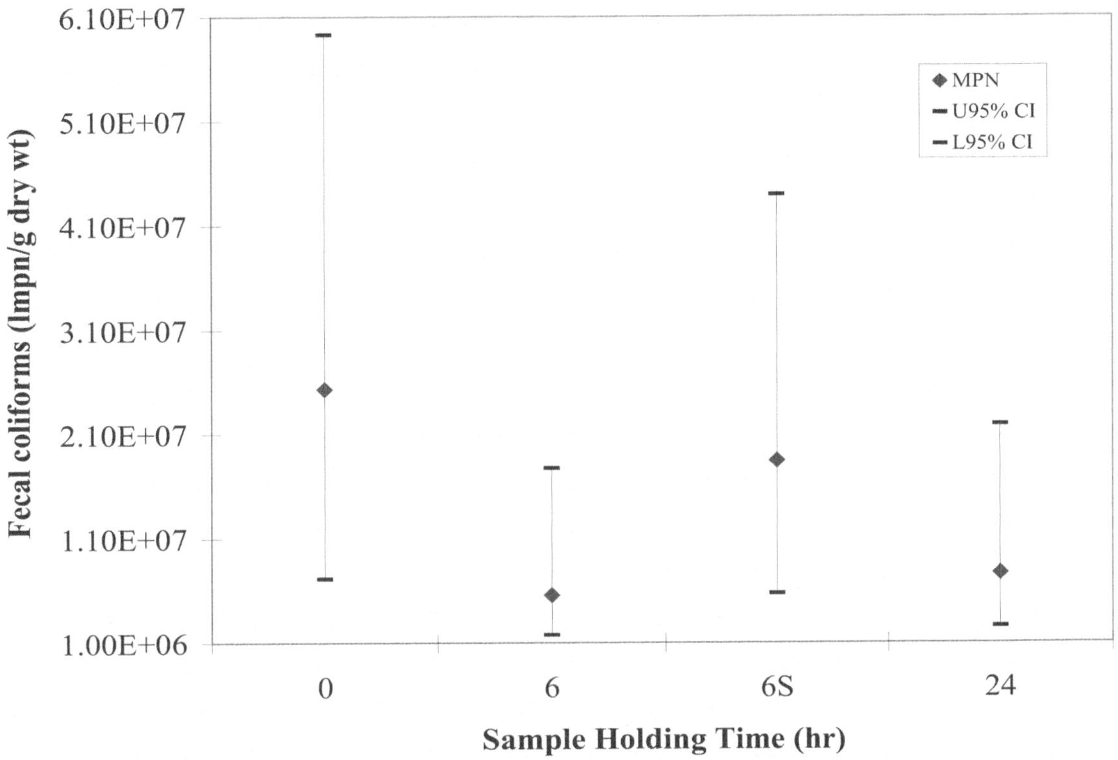

6 – indicates MPN tubes incubated immediately

6S – indicates MPN tubes that sat at room temperature for 2 hours before incubation

Figure 3-9. Results of Sample Holding Time Experiment for Facility #1 Liquid.

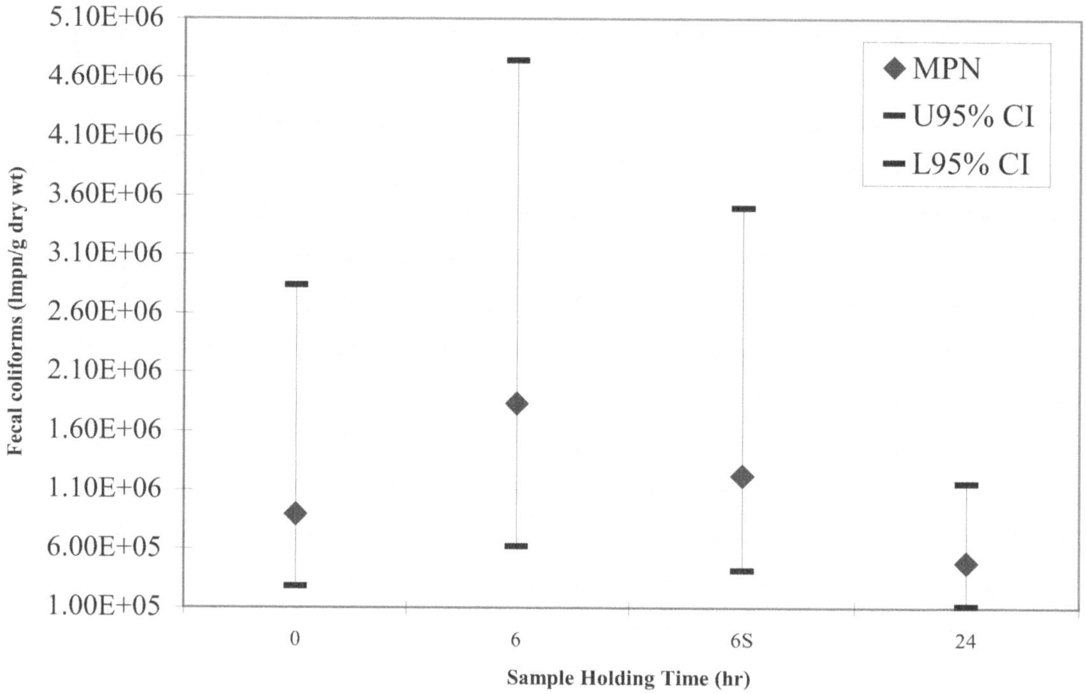

6 – indicates MPN tubes incubated immediately
6S – indicates MPN tubes that sat at room temperature for 2 hours before incubation

Figure 3-10. Results of Sample Holding Time Experiment for Facility #1 Cake.

This experiment was repeated for liquid and cake samples collected from Facility #2. Samples were processed immediately after collection (0 hours), after return to the research laboratory in a cooler (6 hours), and after storage in a cooler overnight (24 hours). For this series of experiments, the tubes inoculated at time 0 were placed in a 35°C portable incubator and incubated immediately. Figure 3-11 presents the results for the liquid samples, and Figure 3-12 presents the results for the cake samples. For Facility #2, liquid samples held for up to 24 hours in a cooler between 4 and 10°C resulted in $3.46x10^7$, $2.32x10^7$, and $5.47x10^7$ MPN/g dry weight for zero, 6 and 24 hours of storage, respectively, with significant overlap of their 95% confidence limits. Storage in a cooler for 24 hours, however, did appear to potentially affect the numbers of fecal coliforms enumerated from Facility #2 cake, with lower enumeration densities after 6 and 24 hours. The levels enumerated were $7.13x10^6$, $7.18x10^5$, and $2.06x10^6$ MPN/g dry weight for analysis after 0, 6, and 24 hours of storage in a cooler, respectively. However, the 95% confidence limits significantly overlapped and as such, these differences do not appear to be significant.

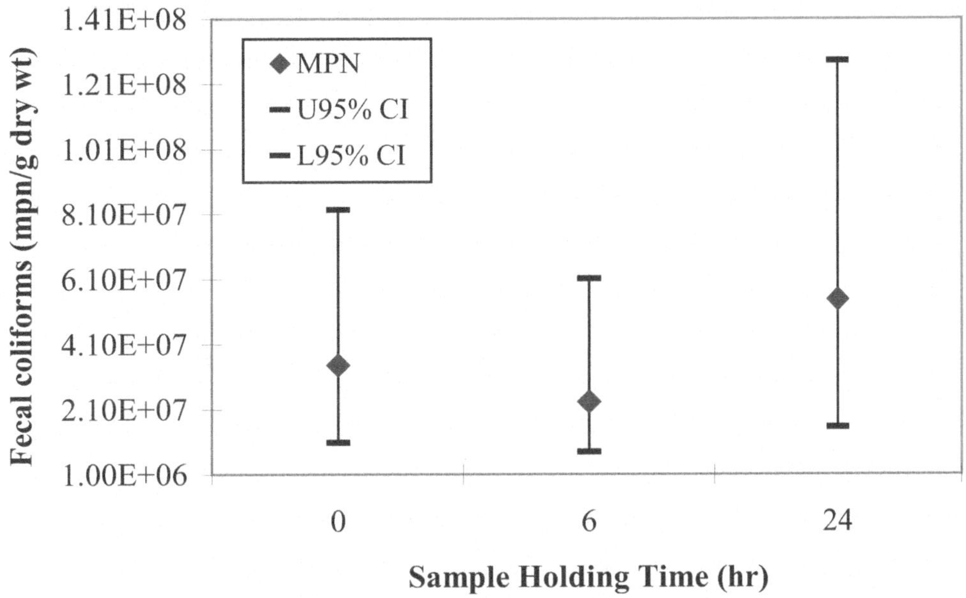

Figure 3-11. Results of Sample Holding Time Experiment for Facility #2 Liquid.

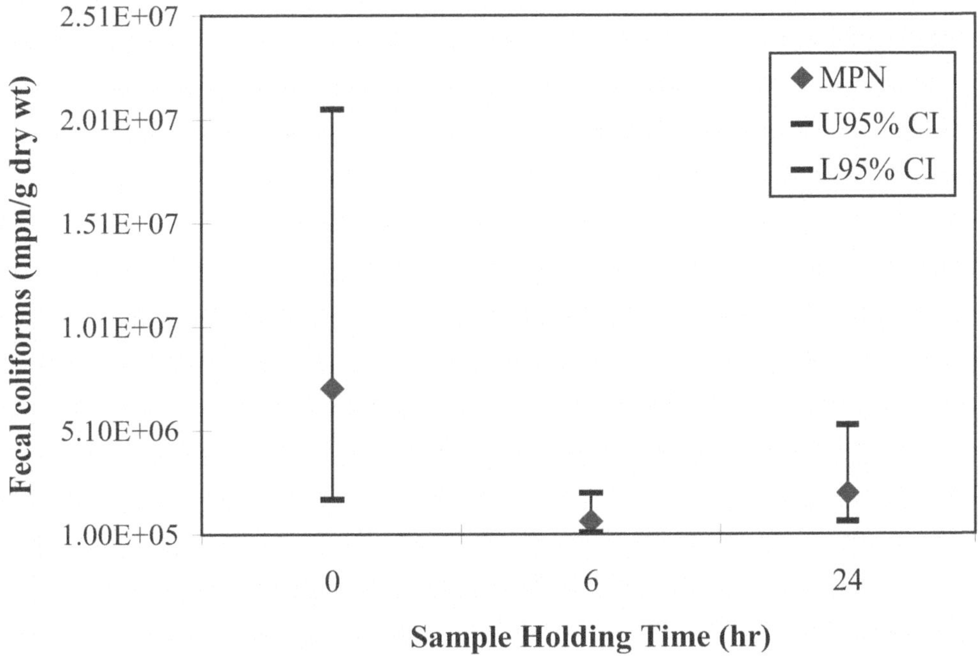

Figure 3-12. Results of Sample Holding Time Experiment for Facility #2 Cake.

The enumeration results from the holding time experiments are summarized in Table 3-8. Qualitatively, all enumerations produced MPN values within an order of magnitude for the same sample. Multiple regression models in SAS were used to assess the significance of holding time on the ultimate MPN per gram dry weight values. Data from both liquid and cake, and each facility were evaluated pooled and separately. The *p*-value for holding time as a variable for the pooled data set was 0.2711, which would allow us to say that holding time was not statistically significant. However, the MSE for this model is 0.0977 and the estimated variance s^2 is 0.0568. Since s^2 is not significantly smaller than MSE, the strength of this conclusion is questionable. When the data for liquid and cake were considered separately, the strength of our conclusion increases slightly, as MSE is approximately four times larger than s^2. It is important to note that the type of material (cake versus liquid) was determined to be significant in determining final MPN value. Again, this lends credibility to our statistical model and conclusions.

Table 3-8. Summary of Sample Holding Time Experiment Data.

Facility #1 liquid (holding time)	MPN/g dry wt	95% CI
0-1 hr	$2.5x10^7$	$7.2x10^6 – 5.9x10^7$
6 hr	$5.6x10^6$	$1.8x10^6 – 1.8x10^7$
6 hr (2 hr RT)	$1.8x10^7$	$5.8x10^6 – 4.4x10^7$
24 hr	$7.7x10^6$	$2.6x10^6 – 2.2x10^7$
Facility #2 liquid (holding time)		
0-1 hr	$3.5x10^7$	$1.1x10^7 – 8.2x10^7$
6 hr	$2.3x10^7$	$7.9x10^6 – 6.1x10^7$
24 hr	$5.5x10^7$	$1.6x10^7 – 1.3x10^8$
Facility #1 cake (holding time)		
0-1 hr	$8.9x10^5$	$2.8x10^5 – 2.8x10^6$
6 hr	$1.8x10^6$	$6.2x10^5 – 4.7x10^6$
6 hr (2 hr RT)	$1.2x10^6$	$4.2x10^5 – 3.5x10^6$
24 hr	$4.8x10^5$	$1.1x10^5 – 1.2x10^6$
Facility #2 cake (holding time)		
0-1 hr	$7.1x10^6$	$1.8x10^6 – 2.1x10^7$
6 hr	$7.2x10^5$	$1.8x10^5 – 2.1x10^6$
24 hr	$2.1x10^6$	$7.0x10^5 – 5.3x10^6$

Additional experiments were conducted to evaluate the effect of holding times on liquid samples. A 1L sample of biosolids was collected from the gravity belt thickener at Facility #4 and two 5x5 sets (10^{-1} through 10^{-5}) of LTB tubes were inoculated in accordance with EPA Method 1680 after each of three holding times: 0 to 1 hour, 6 hours, and 24 hours. In addition, two 5x4 sets (10^{-1} through 10^{-4}) of A-1 tubes were simultaneously inoculated in accordance with EPA Method 1681 at each of the three holding times. Owing to a laboratory error with the A-1 tubes, the six-hour sample results were not usable, and a second experiment was performed in which two holding times were evaluated, 0-1 hour and 6 hours. Samples were analyzed at each holding time, using the same two methods indicated previously, Methods 1680 and 1681. In between time points, the biosolids were stored in a cooler with ice packs and temperature was monitored (see Section 3.3.6 – Cooling Regimes). The results are presented graphically in Figures 3-13 to 3-16. The enumeration results are also summarized in tabular form in Table 3-13.

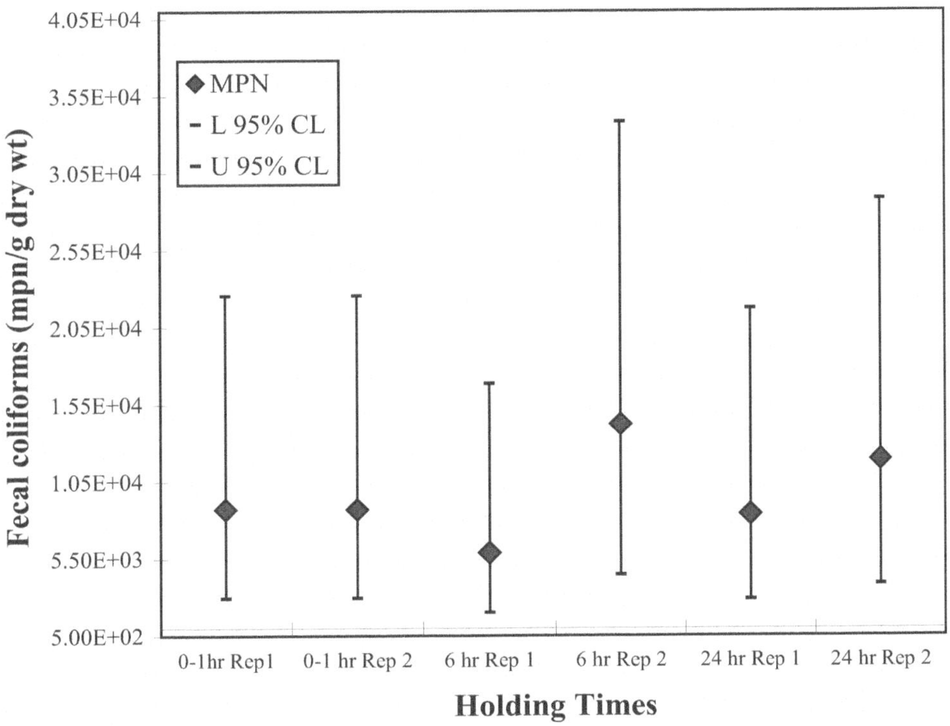

Figure 3-13. Results of Facility #4 Sample Holding Time Experiment Method 1680.

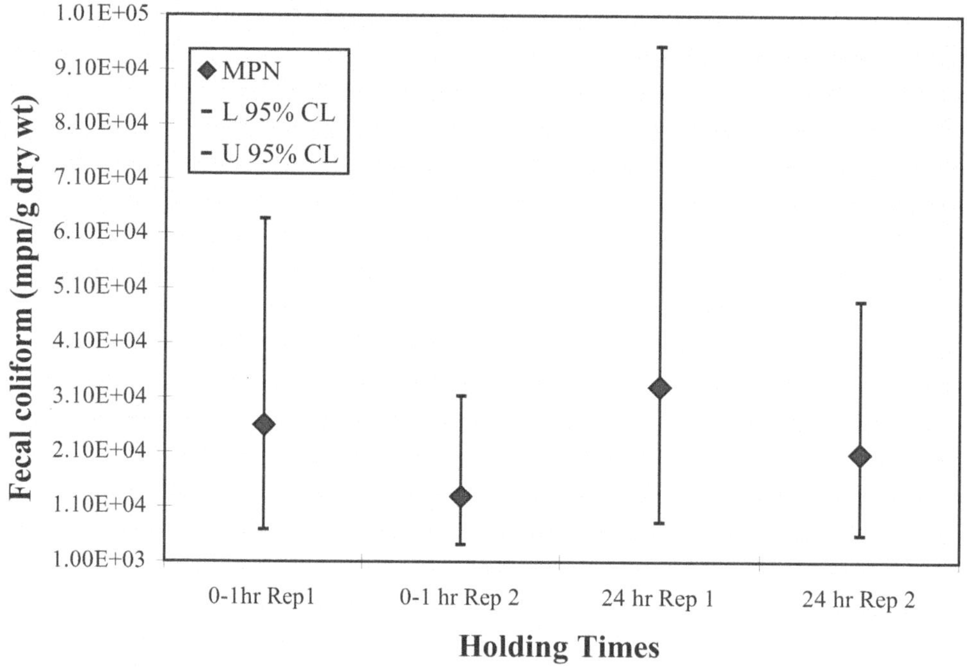

Figure 3-14. Results of Facility #4 Sample Holding Time Experiment Method 1680; 0-1 hr vs. 24 hrs.

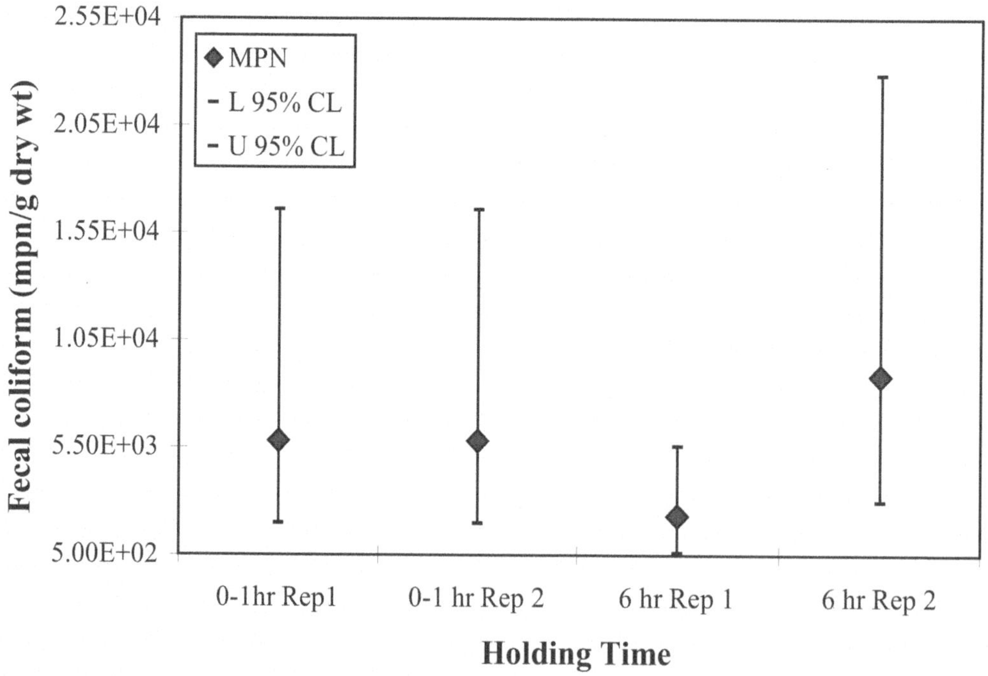

Figure 3-15. Results of Facility #4 Sample Holding Time Experiment Method 1681; 0-1 hr vs. 6 hrs.

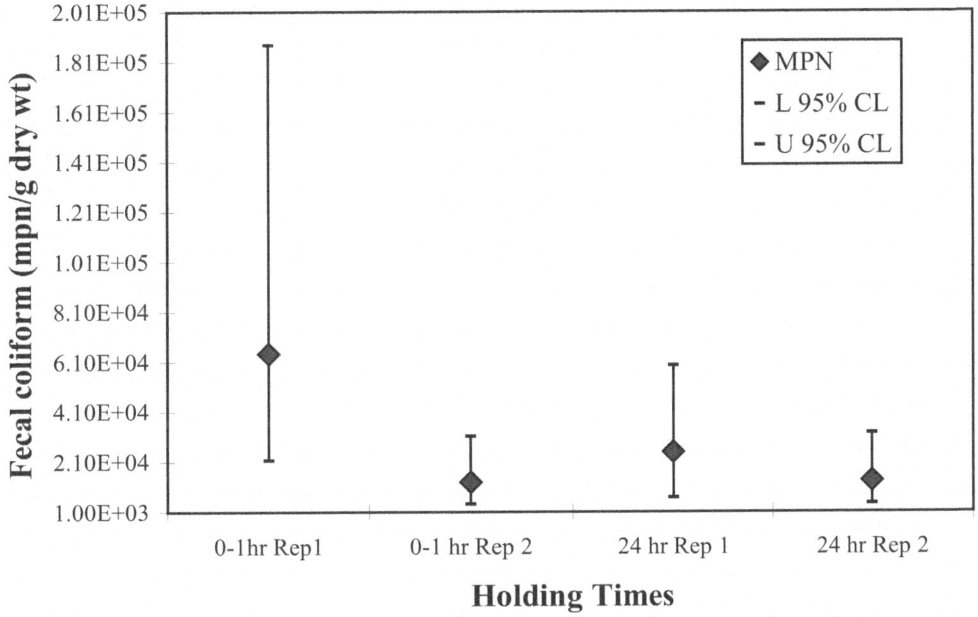

Figure 3-16. Results of Facility #4 Sample Holding Time Experiment Method 1681; 0-1 hr vs. 24 hrs.

Table 3-9. Summary of Facility #4 Sample Holding Time Experiment Data,

Holding Time	MPN/g dry weight	95% CI
Method 1680		
0-1 hr Rep 1	$8.71x10^3$	$2.95x10^3 – 2.25x10^4$
0-1 hr Rep 2	$8.71x10^3$	$2.95x10^3 – 2.25x10^4$
6 hr Rep 1	$5.90x10^3$	$2.01x10^3 – 1.69x10^4$
6 hr Rep 2	$1.42x10^4$	$4.43x10^3 – 3.38x10^4$
24 hr Rep 1	$8.38x10^3$	$2.84x10^3 – 2.17x10^4$
24 hr Rep 2	$1.19x10^4$	$3.81x10^3 – 2.88x10^4$
0-1 hr Rep 1	$2.60x10^4$	$6.80x10^3 – 6.37x10^4$
0-1 hr Rep 2	$1.29x10^4$	$4.14x10^3 – 3.13x10^4$
24 hr Rep 1	$3.30x10^4$	$8.22x10^3 – 9.53x10^4$
24 hr Rep 2	$2.08x10^4$	$5.90x10^3 – 4.87x10^4$
Method 1681		
0-1 hr Rep 1	$5.81x10^3$	$1.98x10^3$ - $1.66x10^4$
0-1 hr Rep 2	$5.81x10^3$	$1.98x10^3$ - $1.66x10^4$
6 hr Rep 1	$2.33x10^3$	$6.24x10^2 – 5.57x10^3$
6 hr Rep 2	$8.84x10^3$	$2.99x10^3 – 2.29x10^4$
0-1 hr Rep 1	$6.43x10^4$	$2.17x10^4 – 1.88x10^5$
0-1 hr Rep 2	$1.29x10^4$	$4.14x10^3 – 3.13x10^4$
24 hr Rep 1	$2.49x10^4$	$6.67x10^3 – 5.96x10^4$
24 hr Rep 2	$1.34x10^4$	$4.29x10^3 – 3.25x10^4$

With the exception of one value enumerated after 24 hours of sample storage, each sample value was within the 95% confidence intervals of all other enumeration values for Method 1680 tests. However, the most-probable-number values enumerated using Method 1681 were more variable.

The processes that take place during sample holding include: 1) continued decay of stressed or injured organisms; 2) regrowth of organisms after removal of treatment process stress; and/or 3) entering "stasis" because of lower temperatures. From observation of enumeration trends, it is suggested that a holding time up to 24 hours is not extremely important in terms of re-growth, as long as the sample is adequately packaged with ice packs or another approved method of cooling. For all three facilities, applying hypothesis testing (Procedure Mixed in SAS) for liquid samples only did not result in statistically significant differences in enumeration values (p=0.4959) based on holding time. However, the mean estimates from 0 to1 hour, 6 hour, and 24 hour increased. This indicates that potentially small, but not statistically significant, amounts of regrowth may occur. When the same analysis was conducted using the liquid and cake sample datasets pooled together, holding time was again not statistically significant (p=0.1826); however, in this case, no trend in estimate means for analysis times was observed.

Additional statistical analyses comparing results from Methods 1680 and 1681, presented above, on identical blender replicates of sample homogenate were conducted. Using the Signed Rank test (Procedure Univariate in SAS), no statistical difference in method enumeration values was observed (p=0.1885). Additionally, using log transformed data at the 95% confidence level, the variance interval for sample values incorporates the null hypothesis (each method yields the same result). Therefore, we cannot reject the null hypothesis. Utilities deciding which fecal coliform method to use should consider experience with the method and, perhaps, consistency with historical data.

3.3.6 Cooling Regimes

QA/QC guidance (U.S. EPA, 2003) recommends that samples be cooled in a slurry of water and ice or with blue ice until the sample temperature is less than 10°C for analysis within 1 to 6 hours, or 4°C for analysis within 24 hours for any intended analysis except for viruses. Samples for viruses may be frozen. Samples for other microbial analyses should not be frozen. In practice, the cooler temperature is typically monitored in a manner that ensures the integrity and aseptic nature of the samples.

Multiple experiments were conducted to document how quickly the center cores of 1L, 2L, and 4L bottles of biosolids cool when placed in a cooler with blue packs, to verify whether larger sample volumes could be adequately shipped while minimizing re-growth. This is especially significant for wastewater treatment plants needing to ship the biosolids to off-site laboratories for enumeration. The longer the time required for the sample to reach 10 to 4°C, the longer the period for potential regrowth along with the corresponding increased possibility of incorrect or skewed enumeration results. Cooling experiments were conducted using a Traceable® waterproof thermometer with a probe cable so that coolers could remain closed for the duration of the experiments. The thermometer could be read from the outside of the cooler.

A sample of biosolids from the gravity belt thickener at Facility #4 was collected in a bottle, which was immediately placed into a cooler with ice packs. The temperature change of the center core of the sample was recorded over time. Each sample ranged in starting temperature from approximately 39°C to 40°C. The experiment was repeated in duplicate each with 1L, 2L, and 4L bottles. The experiment was repeated again using 1L, 2L, and 4L bottles, but this time the biosolids were allowed to cool to room temperature before being placed in a cooler with ice packs and monitoring temperature changes. Figures 3-17 through 3-19 present the cooling experiment results for thermophilic and ambient temperature biosolids of 1L, 2L, and 4L sizes, respectively.

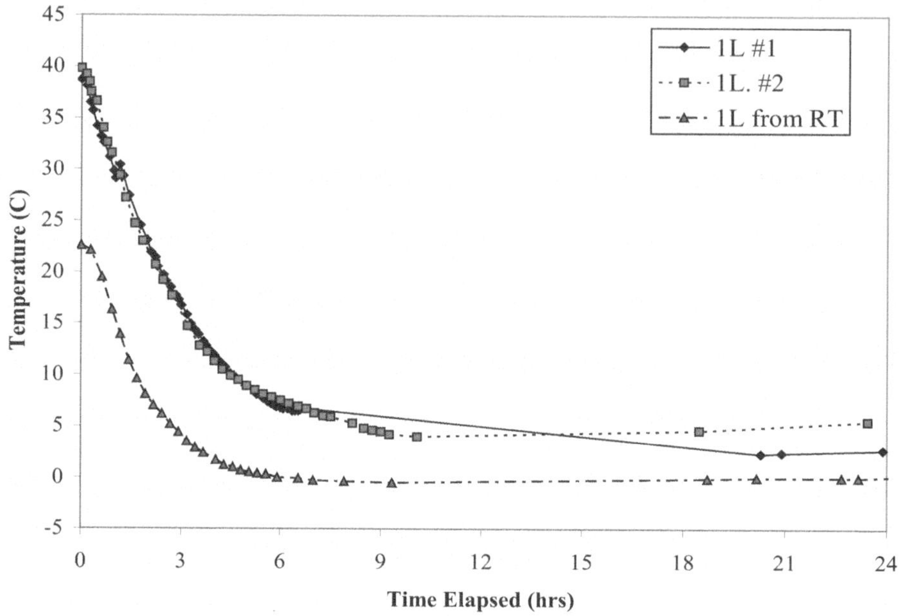

Figure 3-17. Cooling of 1L Samples.

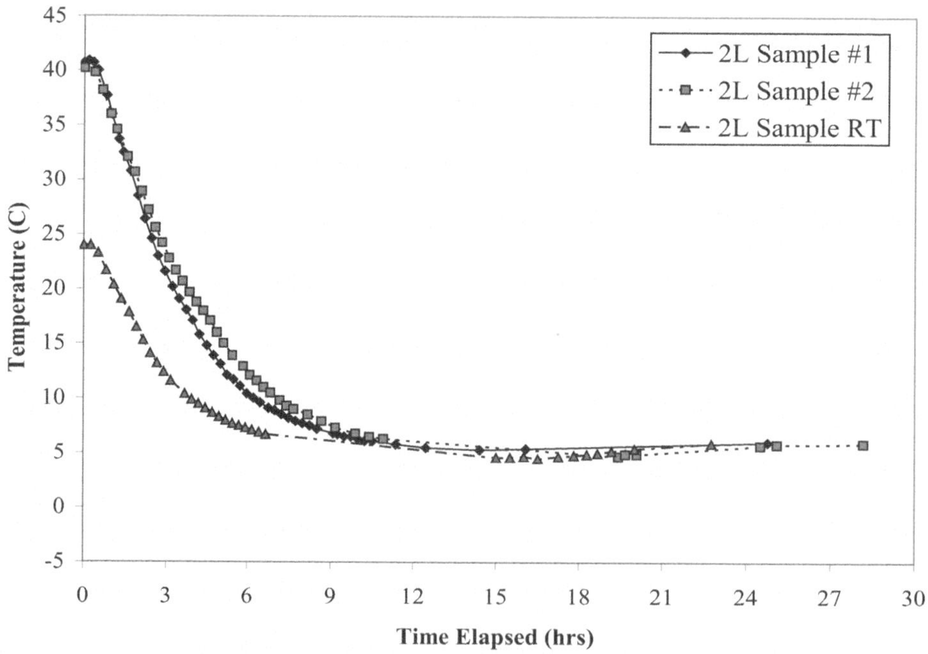

Figure 3-18. Cooling of 2L Samples.

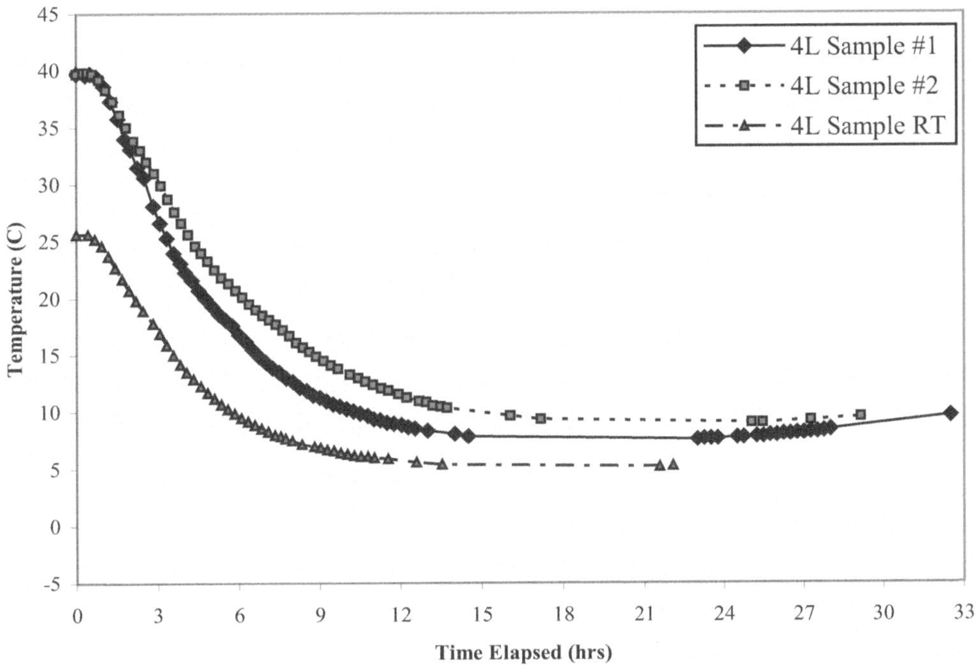

Figure 3-19. Cooling of 4L Samples.

Consistent with EPA guidance (U.S. EPA, 2006, 2006a, and 2006b), 1L samples cool to below 10°C within 4 hours for thermophilic biosolids and in approximately 1 hour and 40 minutes for ambient temperature biosolids. For 2L samples, a 10°C sample core temperature was reached after about 6 hours and 4 hours for thermophilic and room temperature samples, respectively.

The initial temperature of the ice packs may have played the largest part in the experiments involving the 4L bottle. Three cooling experiments using 4L of biosolids were conducted. Because of time constraints, the ice packs used in the second 4L experiment were only stored in the -20°C freezer for approximately 15 hours, whereas in the other cooling experiments the freezer packs were stored for at least 24 hours in the other cooling experiments. This may account for inconsistencies between the cooling curves. Approximately 9.5 hours and 6 hours were required to cool the core of 4L thermophilic and room temperature samples, respectively, to 10°C. The sample cores for the 4L samples never reached below 4°C for the 4L sample volumes during the 22-33 hour monitoring periods. Not surprisingly, collection of 1L samples minimizes the period for possible regrowth during shipping. If biosolids samples larger than 1L are required for analyses, it is recommended that multiple 1L containers be collected and cooled.

There are many variables that should be considered when evaluating the results of these experiments, the most important of which may concern the ice packs used in the cooler. The number of ice packs used and their composition (blue packs or wet ice), the configuration they are arranged in, and how cold the ice packs are at the start of the experiment may all affect the rate of cooling of biosolids samples.

Another variable to consider is the temperature outside of the cooler. If the cooler is stored/transported in a cool environment, the heat transfer between the cooler and the environment should be minimal. Coolers shipped during warm months would be expected to gain heat from the environment to a greater extent than those shipped during cool months. The ice packs may change temperature faster in the warm area because the cooler will gain more heat energy from its surroundings. Therefore, utilities shipping samples off-site during the summer, should appropriately pack the coolers with increased amounts of ice packs.

Another factor that cannot be controlled is the amount of movement or vibration the cooler will experience in transit. The more movement the cooler is subjected to during transportation, the more mixing may occur in the bottle, resulting in faster cooling. More rapid cooling helps to minimize regrowth potential, but is very difficult to achieve without risking freezing the sample.

Finally, other factors may need to be considered including the insulation qualities of the cooler used, as well as the total amount of ice added relative to sample volume. The effectiveness of cooling is strongly rooted in thermodynamics.

3.3.7 Field Sterilization

EPA guidance allows for field sterilization of sampling equipment using a 10% commercial bleach solution followed by thorough rinsing with tap water. However, bleach can react with ammonia, a by-product of biosolids bioactivity. Without a handy drain, disposal of bleach solutions can release this toxic and caustic solution onto the ground. An alternative field cleaning technique that includes three washes with 50% ethanol/sterile water (v/v) followed by three rinses with sterile lab quality water was previously tested (Long, 1986).

Commercially-available blenders are often purchased for use in laboratories because of the cost savings over steel laboratory homogenizers. However, commercial blenders are not autoclavable because their rubber gaskets would degrade. For this experiment, commercially available blenders owned by Facility #1 and the research group were sterilized between samples by rinsing three times with 300mL of 50% ethanol/sterile water (v/v) followed by three rinses with sterile lab quality water. After all samples were processed, the blender was once again washed using this procedure. Then, 270mL of sterile coliform buffer was added to the blender, blended for two minutes, and a series of five 1mL aliquots were inoculated into LTB and, for *Salmonella* analyses, into 3xTSB and 1xTSB tubes. The LTB tubes, termed blender blanks, typically remained negative in all but one experiment. When inoculated with buffer from these blenders, the *Salmonella* enrichment TBS tubes, which contain a general growth media, always remained negative.

Staff at Facility #4 used a plexiglass mixer to stir the biosolids before subsampling. This device is typically soaked in a bleach solution, soap and water rinsed, and allowed to dry between analyses. Just prior to use, the mixer is rinsed with sterile buffered water. QA/QC checks conducted by Facility #4 always remain negative for bacterial growth.

Collectively, these results indicate that if careful guidance for washing procedures for sampling and analysis devices is developed, the need to use a freshly washed, wrapped and autoclaved piece of equipment for each sample collected may not always be necessary.

3.3.8 Compost and Particle Size Effects

Experiments involving the enumeration of large vs. small particles from compost samples were undertaken to investigate the importance of particle size, a factor potentially affected by specific devices used to collect samples. A number of sampling devices, hand-held corers, trier samplers, thief samplers and even curved bladed scoops have been noted to exclude larger pieces from samples collected. It was noted during the WEFTEC05 workshop that the majority of microorganisms may be associated with finer particles in compost, on a dry weight density basis, as the interior of large particles would more likely consist of bulking agent matrix.

For this series of experiments, large particles that could be separated using sterile forceps and smaller remaining particles were extracted from Facility #3 compost samples, identified as samples #2 and #7 from Table 3-5, and enumerated separately. Note that samples #2 and #7 had previously been enumerated to contain 552 and 599 MPN/g dry wt fecal coliforms, respectively. The results of the particle size experiments are presented in Figure 3-20 and summarized in Table 3-10.

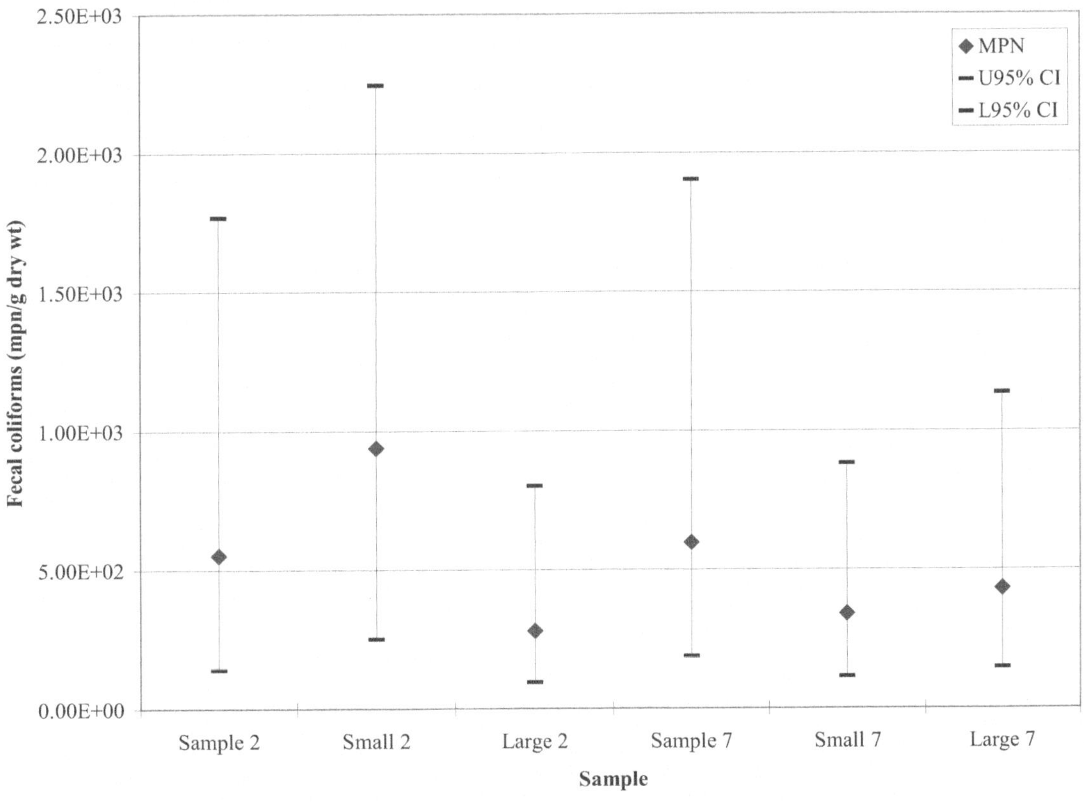

Figure 3-20. Enumeration of Selected Portions of Compost Samples.

Surprisingly, approximately equal densities of fecal coliforms were contained in the small and large-sized particles of the compost from Facility #3. The 95% confidence limits of each measurement for each sample overlapped the calculated numbers for each of the other two measurements for the other two sample categories. Overall, this indicates that microbial content may not be significantly over-estimated if sampling methods were to exclude large particles as an artifact of sampling device. However, this observation is only specific to the product tested. Other facilities may have less homogenized product where microbial density is more strongly correlated with particle size. Using multiple regression analysis, the *p*-value for particle size as a factor was 0.2174 which indicates that particle size is not significant. However, since the MSE and s^2 were approximately equal (~0.06), the strength of this conclusion is extremely weak.

The relatively common practice of scraping away the top 6 inches of compost before quartering and sampling was noted by PSC members and WEFTEC05 workshop participants as potentially introducing bias into the sampling approach, and *a priori* resulting in collection of a non-representative sample. To address this issue, a pile of compost from Facility #3 was divided into a regular grid. Four samples were collected from the top 6 inches of the pile location, and also after the top 6 inches were scraped away. The sample bottles were placed in a cooler and processed upon return to the laboratory at UMass, 2 to 4 hours following sample collection. The results of these enumerations are presented in Figure 3-21 and summarized in Table 3-10. Not surprisingly, the levels of fecal coliforms present in the surface samples appear to be lower than those from the deeper samples with geometric means of 1.7×10^4 and 6.2×10^4 MPN/g dry weight, respectively. For this sample site, exposure to sunlight and desiccation may play a role in the observed microbial content differences. Hypothesis testing indicates that the geometric means were not significantly different at $\alpha=0.05$, or 95 percent confidence. To analyze these data, a paired t-test was used. The *p*-value was 0.2610 suggesting that scraping away the surface layer does not significantly affect results. However, in each case, the paired surface sample contained a lower density of fecal coliforms than its corresponding deep sample. Once again, the strength of the statistical conclusion is weak due to the large mean of the differences among the data.

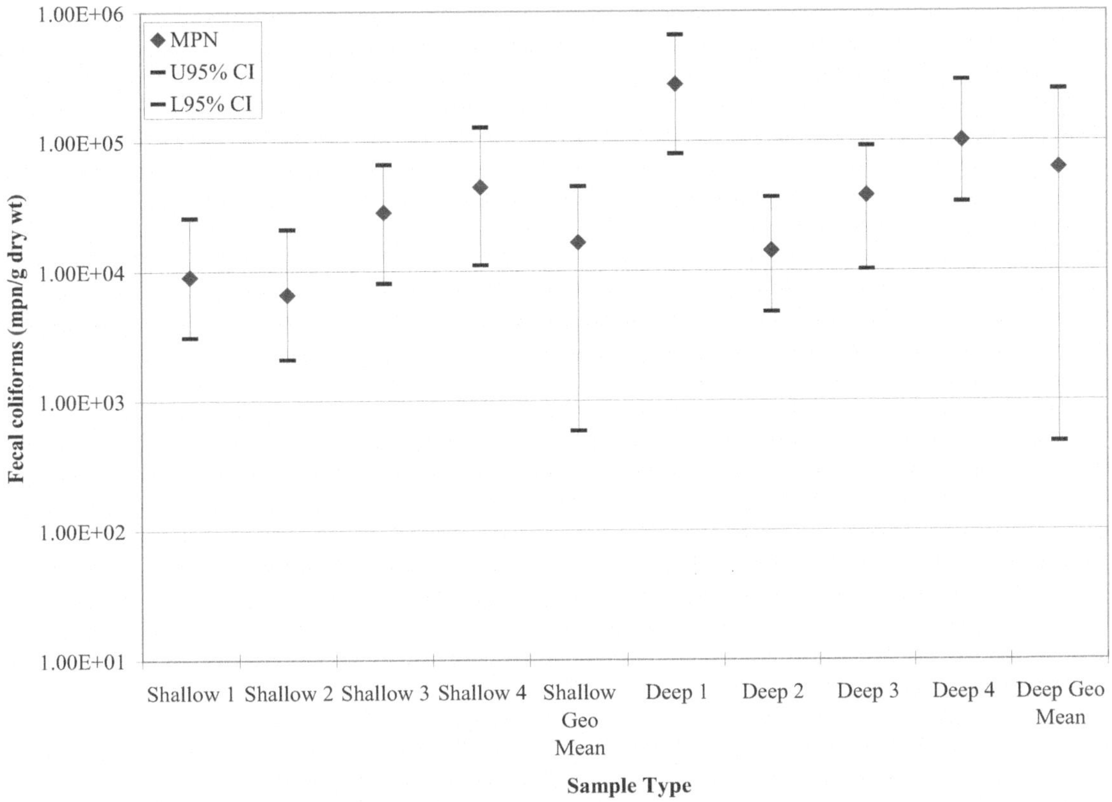

Figure 3-21. Enumeration of Surface vs. Deep Compost Samples.

Table 3-10. Results of Particle Size and Pile Depth Compost Sample Analysis.

Sample ID	FC by 1680 (MPN/g dry wt)	95% CI
Small 1	$9.4x10^2$	250 - 2,200
Small 2	$3.4x10^2$	120 - 880
Large 1	$2.8x10^2$	96 - 800
Large 2	$4.3x10^2$	150 - 1,100
Shallow 1	$9.1x10^3$	3,100 - 26,000
Shallow 2	$6.6x10^3$	2,100 - 21,000
Shallow 3	$2.8x10^4$	8,100 - 67,000
Shallow 4	$4.4x10^4$	11,000 - 130,000
Geometric mean	$1.7x10^4$	
Deep 1	$2.7x10^5$	80,000 - 660,000
Deep 2	$1.4x10^4$	4,800 - 37,000
Deep 3	$3.8x10^4$	10,000 - 91,000
Deep 4	$1.0x10^5$	34,000 - 290,000
Geometric mean	$6.2x10^4$	

3.3.9 Multiple Microbial Measures

Two experiments were conducted where samples of anaerobically digested, gravity thickened liquid were split between project personnel, Facility #4 wastewater personnel, and a contract laboratory. As discussed in Section 3.1.4, Facility #4 was in the process of transitioning from a Class B to Class A product. Samples in the first experiment were analyzed for fecal coliform using Methods 1680 and 1681, *Salmonella* using Method 1682, and coliphage using Method 1602. Samples in the second experiment were analyzed for fecal coliforms using Method 1680 and 1681, *Salmonella* using Method 1682, coliphage using Method 1602, total culturable virus, and helminth ova. The results were tabulated and determined to meet Class A criteria for microbial levels (FC < 1000 MPN/g; *Salmonella* < 3 MPN/4 g; coliphage <1 pfu/g; and total virus <1/4 g). The results are summarized in Tables 3-11 and 3-12.

Table 3-11. Multiple Measures Experiment 1.

	Project Team								Wisconsin Utility		
	1680		1681							1680	
Sample	FC	A/U	FC	A/U	Sal	A/U	Phage	A/U	Sample	FC	A/U
1	870	A	387	A	5.2	U	ND	A	1	717	A
2	1320	U	2070	U	ND	A	65	U	1a*	426	A
3	774	A	349	A	ND	A	ND	A			
4	1530	U					ND	A			
5	349	A					ND	A			
6	870	A					ND	A			
7	2070	U					ND	A			

FC - fecal coliforms
Sal - *Salmonella*
A - acceptable for Class A
U - unacceptable for Class A
ND - not detected
*split sample from same 1L

Table 3-12. Multiple Measures Experiment 2.

	Project Team								Wisconsin Utility			Contract Lab		
	1680		1681							1680				
Sample	FC	A/U	FC	A/U	Sal	A/U	P	A/U	Sample	FC	A/U	Sample	Virus	A/U
1	5060	U	7580	U	ND	A	169	U	1	2000	U	1	<1.25	U
1a	2650	U	7020	U	ND	A	177	U	2	4620	U	2	<1	A
2	5010	U	2630	U	ND	A	130	U	3	700	A	3	<1	A
2a	7510	U	3660	U	ND	A	61	U						
3	2590	U	147000	U	ND	A	86	U						
3a	2730	U	2020	U	ND	A	125	U						

FC - fecal coliforms
Sal - *Salmonella*
P - phage
A - acceptable for Class A
U - unacceptable for Class A
ND - not detected
#a - replicate from same 1L sample

The data demonstrate that, for instances where microbial content is close to the Class A regulatory compliance level, analysis of multiple individual discrete samples is warranted. In this case, Facility #4 was transitioning to Class A while distribution of biosolids was continuing under Class B provisions. For the first experiment, when fecal coliform levels were very close to the Class A compliance level, three discrete samples analyzed by any of the methods tested were enough to capture the unacceptable levels. For the second experiment, where fecal coliform levels were substantially greater than the Class A compliance level, three samples were enough to capture unacceptable fecal coliform levels; however, all *Salmonella* results met Class A criteria. For facilities with good *Salmonella* destruction, measuring fecal coliform may be a conservative indicator. In other words, measuring fecal coliforms would indicate potential pathogen presence when the pathogen, *Salmonella*, is in fact absent (or below detection limits).

CHAPTER 4.0

FIELD STUDIES

The field study phase (Phase III) of this project involved evaluating sludge and biosolids sampling approaches in treatment plants. A member of the project team traveled to participating utilities and conducted sampling after discussions with facility staff at intensive study sites. Samples were collected and analyzed for fecal coliform using EPA Method 1680. The utilities' standard sampling procedures were discussed. Historical data were requested, in order to review the variability and consistency in testing results. Two additional utilities were interviewed (brief study sites). Qualitative and quantitative evaluations of the resulting Phase III datasets were used to help develop recommended sampling approaches for liquid, cake and/or compost based on qualitative and quantitative evaluations of the resulting datasets. The characteristics of the Phase III facilities are briefly summarized in Tables 4-1 and 4-2, with additional details provided in subsequent sections on each participating facility.

Table 4-1. Summary of Intensive Study Phase III Utilities.

Facility ID	Design Flow	Biosolids Product	Primary Liquid/Solid Treatment Processes
SE Utility 1	29 mgd	Class B liquid	Activated sludge, biological P removal, anaerobic digestion
SE Utility 2	12 mgd (upgrading to 14.5 mgd)	Class B and Class A liquid	Activated sludge, BNR Process operated mainly for P removal, anaerobic digestion in series
MW Utility 1	3.1 mgd	Class B liquid	Activated sludge, chemical P removal, anaerobic digestion in series
MW Utility 2	2.6 mgd	Class A cake	Activated sludge, chemical P removal, TPAD
MW Utility 3	48 mgd	Class B cake	Pure oxygen activated sludge, multimedia filters, TPAD, centrifuge dewatering
NW Utility 1	133 mgd (design ave. wet weather flow)	Class B cake	Activated sludge, anaerobic digestion, high solids centrifuge dewatering
SE Utility 3 (biosolids only)	75,000 tons annually (Contractor)	Class A compost	Aerated static pile with proprietary bulking agent mix followed by trommel separation and curing (various times)
SE Utility 4	14.2 mgd	Class A compost	Advanced biological treatment, in vessel, woodchip composting followed by curing and trommel screen separation
NE Utility 1	2.2 mgd	Class A compost	Activated sludge with biological phosphorus removal. Aerated static pile, anaerobically digested primary solids and WAS blend and woodchip composting followed by trammel separation and curing (at least 52 days)

Table 4-2. Summary of Brief Study Phase III Utilities.

SW Utility 1	18 mgd	Class A liquid	Activated sludge, anaerobic digestion, lagoon storage
NW Utility 2	115 mgd (design average wet weather flow)	Class B cake	Activated sludge, anaerobic digestion, high solids centrifuge dewatering

4.1 Field Study Results

4.1.1 SE Utility 1 – Class B Liquid

This facility produces their biosolids product by anaerobically digesting a blend of waste activated sludge (WAS) and primary solids in a continuous flow process. The solids are transferred into the anaerobic digester using a step-feed process where the target tank temperature is 37.8°C (100°F) with a solids residence time (SRT) of 28-32 days. The temperature in the anaerobic digester is monitored 24/7 by a PLC-based plant control system. During the step-feed process, the finished product is pumped to a gravity belt thickener (GBT) and fed into holding tanks. A total of four open holding tanks fitted with impeller mixers and a capacity of 2.8 million liters (750,000 gallons) are utilized. An additional two tanks are available if additional storage capacity of finished liquid biosolids is required.

SE Utility 1 produced approximately 3,500 dry tons of biosolids in 2006, all of which was land applied for agricultural uses. Based on this production rate, the facility is required to analyze seven biosolids samples every 60 days for microbial content and report the geometric mean. This facility uses an algorithm to distribute sample collection within the 60-day window to generate results that are representative of the product leaving the plant. Samples intended for microbial analyses are collected from storage tanks while they are being mixed. The mixers are turned on at least one hour before samples are to be collected, or liquid land application trucks are to be filled. A length of cotton twine is used to lower a sterile glass bottle to near the bottom of the tank, and the bottle is slowly raised to the surface to collect a grab sample. A sterile specimen jar (approx. 150 mL) is filled by mixing grab samples from several locations around the perimeter of the tank. Once full, the specimen jar is placed in a cooler, and delivered to the analytical laboratory. Transport times typically result in sample analysis within 6 hours.

For this project, seven 1L samples were collected from different locations around the perimeter of the tank on May 31, 2007. Sterile, wide-mouth, plastic bottles were filled to 0.5 to 1 inch of the rim using the grab method described above. Each bottle was filled from one or two grab samples taken at a given location. Most samples were taken during active mixing, except for one sample that was taken next to a mixer impeller. For this sample, the mixer was turned off briefly while the sample was taken to avoid breaking the glass sample bottle; however, the liquid was still well-mixed. Samples were placed on ice immediately and analyzed using EPA Method 1680 within 3 hours of collection. For comparison purposes, the range and variation in the seven samples collected as part of this project were compared to the range and variation reported for the last four series of samples collected by the utility. This data is summarized in Table 4-3.

For all seven samples collected in a single day, the coefficient of variation was 29%. For samples collected by the utility (typically three samples on one day and four on another), the coefficient of variation ranged from 66% to 293% for the four sampling campaigns reviewed. The use of a sampling scheme on multiple days captures a greater amount of variability in the biosolids product when compared to collecting samples all in a single day from a well mixed storage tank. Since finished liquid biosolids are added to the storage tanks daily, sampling on multiple days captures more inter-batch process variability. On no occasion did any of the samples exceed the Class B monitoring limit of 2 million MPN fecal coliforms per gram dry weight.

Table 4-3. Comparison of Fecal Coliform Analyses of SE Utility 1 Samples.

Sample	WERF (MPN/g dry wt)	Utility (1/10 & 1/18/07)	Utility (2/14 & 3/7/07)	Utility (5/1 & 6/6/07)	Utility (4/10 & 4/11/07)
1	187,000	29,000	14,200	28,600	7,660
2	105,000	7,250	14,100	8,130	7,810
3	114,000	7,220	24,600	8,100	15,400
4	168,000	80,400	60,900	95,500	30,200
5	170,000	59,000	46,900	274,000	29,000
6	293,000	57,000	46,600	870,000	7,380
7	119,000	32,000	107,000	196,000	51,500
Geo mean	156,000	27,900	35,400	70,200	16,300
Coeff. Var.	29%	82%t	66%	293%	82%

4.1.2 SE Utility 2 – Class A and Class B Liquid

The liquid treatment side of this facility consists of conventional preliminary treatment (screening, degritting, flow measurement), primary settling followed by an activated sludge process set up for biological nutrient removal, mainly phosphorus removal. The phosphorus removal process utilizes an unaerated anaerobic zone through which all of the return activated sludge (RAS) is passed. In this unaerated zone, a fermentation product liquor is added to provide the necessary food to support the phosphorus removal process. The fermentation process liquor is raw primary sludge that has been held in an anaerobic digester with a very short detention time, creating the necessary volatile fatty acids for P removal. Secondary clarifier effluent is additionally treated using deep bed filters (which can be utilized for denitrification) followed by UV disinfection and post aeration. Because of the high degree of treatment, there are plans for the plant effluent to be reused mainly in evaporative cooling towers at a nearby state university. In general, the processes following the activated sludge process contribute little solids load in this facility.

This facility produces their biosolids product by anaerobically digesting a blend of thickened waste activated sludge (WAS) and primary fermented solids in a four-stage process. Four anaerobic digesters are operated in series with three being operated in the thermophilic temperature range and one in the mesophilic temperature range. The overall goal is to produce a biosolids product that meets the EPA 503 Class A requirements for anaerobically digested liquid sludge, and specifically Equation 3 of paragraph 503.32-3ii.D of the rule. The primary digester is fed on a near-continuous basis and the temperature is maintained at around 55.6°C (132°F). The average hydraulic detention time for the sludge solids in this first stage for calendar year 2006 was approximately 11.5 days. This digester is mixed continuously utilizing a compressed gas mixing system discharging sequentially through lances arranged around the perimeter of the digester roof. The second stage digester is where the EPA 503 Class A detention time and temperature requirements are met. This digester temperature is maintained at approximately

55.6°C (132°F) and the solids are transferred in small intervals so there is at least a 22-hour holding time (typically longer) in this tank to meet the 503 criteria for pathogen reduction. The average hydraulic retention time in this tank for 2006 was 5.7 days. This digester is mixed continuously, utilizing a compressed gas mixing system discharged through lances arranged around the perimeter of the digester roof. The third stage of the digestion process receives the biosolids in daily batches that have met the Class A requirements in the second stage. For calendar year 2006, the average hydraulic detention time in the third-stage tank was approximately 5.0 days. From the third stage, the biosolids are transferred on a near continuous basis to the fourth stage. This tank is mixed continuously utilizing two draft tube type mixers mounted 180 degrees apart from each other. The fourth stage is operated at mesophilic temperatures to achieve additional volatile solids reduction, to enhabce gas production and to reduce the odor from the thermophilically digested biosolids. For calendar year 2006, the average hydraulic detention time in this tank was approximately 13.8 days. This tank is mixed continuously utilizing two draft tube type mixers mounted 180° apart from each other. From the fourth stage digester, the biosolids are pumped to two covered holding tanks that were originally designed to be aerobic digesters but were never utilized as such. The capacity of these two in-plant storage tanks is approximately 2.3 million liters (630,000 gallons). Not counting the in-plant storage tanks, the overall average hydraulic detention time through the four anaerobic digesters for calendar year 2006 was approximately 36 days.

SE Utility 2 produces approximately 1,800 dry tons of biosolids, of which about 85% is land applied for agricultural uses with the rest dewatered on an as-needed basis and transported to a local commercial composting facility. Based on this production rate, this facility is required to analyze biosolids samples bimonthly. During each sampling period, samples are analyzed on eight consecutive days. Since the plant was undergoing an upgrade from 12 mgd to 14.5 mgd and transitioning to a Class A product, the facility had been collecting and analyzing additional samples for fecal coliform each Monday, Tuesday and Wednesday since January 2007. This facility uses aerators to mix their tanks for at least one hour before sample collection. A Dempster-brand hand pump is located in each storage tank with the bottom of the intake located approximately 3' off the bottom of the tank. The pump is pumped 40 times to clear the pipe prior to filling the sample jars. Dipper samplers are also available if samples from near the surface of the tank are to be collected when the storage tank is near capacity. Samples for regulatory compliance monitoring are collected in sterilized, 1,000mL, glass sample jars. Supplemental monitoring samples are collected in 150mL sterile specimen cups. The samples are delivered to an in-house laboratory for analysis, requiring transport across a distance of only approximately 250'.

For this project, seven 1L samples were collected from the treated biosolids holding tank on June 4, 2007. Sterile, wide-mouth, plastic bottles were filled to 0.5 to 1 inch of the rim using the hand pump after mixing the tanks for approximately an hour via aeration. The utility collected a 150 mL grab sample at the same time. The sample bottles were placed on ice immediately. The samples were analyzed using EPA Method 1680, and analyses were completed within three hours of collection. The utility analyzed their sample using EPA Method 1681. For comparison purposes, the results of the grab samples collected and analyzed by the utility the week of June 4th are presented. The data are summarized in Table 4-4 along with the project team results. Maintaining biosolids temperatures at 55.6°C (132°F) for a minimum of 22 hours to 5.7 days produces a very consistent product. The combination of mixing the storage tanks and collecting one sample per day for several days appears to capture the variability of the biosolids produced by this facility.

Table 4-4. Comparison of Fecal Coliform Analyses of SE Utility 2 Samples.

Sample	WERF (MPN/g dry wt)	Utility grab (6/4/07)	Utility grab (6/5/07)	Utility grab (6/6/05)
1	<70	15	26	27
2	<70			
3	<70			
4	<69			
5	<67			
6	<67			
7	<67			
Geo mean	<69			22
Coeff. Var.	2%			23

4.1.3 MW Utility 1 – Class B Liquid

This facility produces their biosolids product by anaerobically digesting belt thickened WAS in a step-feed process where the solids are step-fed into a heated anaerobic digester with a target tank temperature of 37.8°C (100°F) and a solids residence time (SRT) of 30 to 120 days. The plant maintains two digesters in series; however, the secondary digester is not heated. The mean solids residence time varies because the biosolids are only land applied during a limited time window in the spring and in the fall. The temperature in the anaerobic digester is monitored daily using a thermometer on the heat exchanger. During the step-feed process, the finished product is pumped to a gravity belt thickener (GBT) and fed into holding tanks. A total of two covered holding tanks with a capacity of approximately 2.8 million liters (750,000 gallons) are utilized.

MW Utility 1 is a small treatment plant; the biosolids are held in the two storage tanks and typically land applied only in the spring. On occasion they do some land application during the fall months, but not every year. When the project team sampled, the liquid levels in the storage tanks were relatively low and the material relatively new. Approximately 400,000 gallons of biosolids were produced in spring 2006 for land application for agricultural uses. Based on this production rate, this facility is required to analyze seven biosolids samples once a year for microbial content and report the geometric mean. This facility analyzes one grab sample each day for seven consecutive days, once the spring growing season commences, to achieve results more representative of the product leaving the plant. Samples intended for microbial analyses are collected from storage tanks via a side spigot while pump mixing is engaged, or from a side spigot in the truck loading area. A sterile sample jar (500mL) is filled by placing it beneath the sample spigot. Once full, the specimen jar is delivered to the onsite analytical laboratory for analysis using the LTB/EC multiple tube fermentation method.

For this project, seven 1L samples were collected from the side spigot of a mixing pump in the first holding tank on June 20, 2007. Sterile, wide-mouth, plastic bottles were filled to 0.5 to 1 inch of the rim after the side spigot had been purged of several liters of material. The sample bottles were placed on ice immediately. The samples were analyzed using EPA Method 1680, and analyses were completed within three hours of collection. For comparison purposes, the range and variation in the seven samples collected as part of this project were compared to the range and variation reported for the last four series of samples collected by the utility. These data are summarized in Table 4-5.

The numbers enumerated by the project team are three to four orders of magnitude lower than those enumerated by the utility. A potentially significant difference between samples is that the project team collected samples of relatively "new" biosolids. The utility typically samples the material in their storage tanks just prior to the spring application season and after material has been added over the winter. The tanks were nearly emptied by the time the project team sampled in mid-May with approximately 400,000 gallons having been land applied for agricultural uses. One storage tank was at approximately one-third capacity on the date of sampling with a mean solids residence time of approximately three weeks. This younger biosolids product was also more highly variable than the material sampled by the utility, although the numbers are influenced by one apparent outlier in the project team's dataset. In addition, the data indicate that significant indicator regrowth occurs during the winter and early spring storage period.

Table 4-5. Comparison of Fecal Coliform Analyses of MW Utility 1 Samples.

Sample	WERF (MPN/g dry wt)	Utility (3/20-26/07)	Utility (4/20-24, 5/2, & 5/3/06)	Utility (4/19-26, 5/3/05)	Utility (4/23-25, 5/4 & 5/11/04)
1	430	1,021,277	680,851	1,021,27	3,720,930
2	475	1,120,732	627,907	666,667	697,674
3	<431	2,222,222	627,907	1,043,478	810,811
4	965	3,720,930	454,545	1,043,478	555,556
5	483	3,902,439	454,545	1,066,667	571,429
6	5731	3,809,524	925,926	1,043,478	395,349
7	482	2,325,581	326,923	200,000	309,524
Geo mean	733	2,290,000	558,000	773,000	699,000
Coeff. Var.	173%	46%	27%	32%	111%

4.1.4 MW Utility 2 – Class A Cake

The service area for this facility is mostly residential with a permanent population of about 9,500 and a large summer visitor base. This facility also accepts about 28 million gallons per year of septage from septic system haulers. During the peak season, this can amount to five million gallons per month accounting for approximately 6-7% of the plant flow. This facility produces their biosolids product by anaerobically digesting a mixture of primary solids and WAS in a hybrid feed process. The solids are step-fed into the anaerobic digester hourly for 12 hours, and allowed to mix and react for three hours. Then, a volume equal to the feed volume is transferred to the secondary reactor. The primary reactor has a capacity of 350,000 gallons. The target temperature in the primary reactor is 55°C (131°F) with a theoretical solids residence time (SRT) of about 20 days at a feed volume of about 18,000 gallons per day. The plant maintains two digesters in series, with the secondary digester maintained at a target temperature of 35°C (95°F). The solids are dewatered after secondary digestion using a belt filter press and then are conveyed directly into trucks where it is subsequently delivered for landscape uses. The end-user applies the material to land that is used to grow nursery trees. The field is treated and allowed to sit fallow for about a year. Saplings are planted and grown for about four years, after which they are sold for landscape uses.

All of the biosolids produced by MW Utility 2, approximately 380 dry tons, was used in tree nurseries. The facility samples effluent solids from the primary digester, secondary digester, and dewatered cake weekly for fecal coliforms. The samples are collected in late morning, shortly before the belt press is shut down. The samples are collected in sterile 250mL bottles or 150mL specimen cups, iced and packaged for shipment to the contract laboratory for fecal coliform analysis. The samples are shipped via UPS, and testing is started within 24 hours of collection. The most-probable-number method is used by the analytical laboratory.

For this project, seven 1L samples were collected from the belt filter press conveyer on August 20, 2007. Sterile, wide-mouth, plastic bottles were filled to 0.5 to 1 inch of the rim by scooping cake with the bottle from the top of the material on the conveyer without touching the belt. The samples were collected five minutes apart and placed on ice immediately. The samples were analyzed using EPA Method 1680, and analyses were completed within two hours of collection. For comparison purposes, the range and variation in the seven samples collected as part of this project were compared to the range and variation reported for samples collected by the utility during the preceding three quarters. Overall, the facility produces a very consistent final product with levels of fecal coliforms in the teens per gram dry weight solids in the biosolids cake. Single day sample analyses conducted by the project team also confirmed that the levels of fecal coliforms in this biosolids cake ranged from <9 to 54 MPN/g dry wt, with a geometric mean of 18. This number is a bit higher than those measured by the utility; however, well below the Class A standard in each case. Weekly monitoring exceeds the frequency required by the 503 regulations, but allows the utility the ability to detect process upsets on a near "real-time" basis.

Table 4-6. Comparison of Fecal Coliform Analyses of MW Utility 2 Cake Samples.

Sample	WERF (MPN/g dry wt)	Utility (Jan - Mar 07)	Utility (Apr - Jun 07)	Utility (Jul - Sept 07)
1	10	<12	<11	<13
2	<9	<13	<12	<9.6
3	10	<11	<11	8.9
4	54	<12	<12	<9.1
5	20	<14	<12	<8.6
6	35	<12	<12	<9.9
7	<9	<12	<12	<13
8		<13	12	<12
9			<11	<13
10			<11.3	
11			14	
12			11	
13			12	
Geo mean	18			
Coeff. Var.	81%			

4.1.5 MW Utility 3 – Class B Cake

The service area of this facility includes a combination of residential and industrial customers. The major type of industry served is paper mills, whose wastewater comprises over 50% of the plant's flow. This facility produces their biosolids product by treating secondary solids in a temperature phased anaerobic digestion (TPAD) process. Thickened solids are fed into an anaerobic digester where the target tank temperature is 55°C (131°F) with a solids residence time (SRT) of about seven days. The solids are transferred to a second tank at 40.5°C (105°F) for about 21 days. At the end of the digestion process, the solids are then pumped to two high solids centrifuges for dewatering. The dewatered material is transported on a conveyor belt to a covered concrete storage bay. Fifteen days worth of material can be stored at the treatment facility. The material is loaded directly onto trucks from the storage bay. The utility has an enclosed off-site storage building that is also available and can store one month's production of biosolids.

MW Utility 3 produced approximately 8,743 dry tons of biosolids in 2006 that was land applied for agricultural (10 months per year) and mineland reclamation (two months per year) uses. This facility voluntarily collects and analyzes seven samples once per year. One sample per day is collected over approximately a two to three week period. Each sample is produced by collecting biosolids cake in a sterile beaker and then mixing with a spatula. About three grab samples are composited over a three-minute period and placed in a Whirl-Pak® bag and sent for analysis. The utility uses a contract lab to perform fecal coliform (Method SM9221E) and *Salmonella* (EPA Method 625/R-92/013) analyses.

For this project, seven 1L samples were collected on July 23, 2007. Sterile, wide-mouth, plastic bottles were filled to 0.5 to 1 inch of the rim by dipping the bottle several times into the cake as it passed by on the conveyor. Care was taken not to touch the belt surface. The samples were placed on ice immediately. The samples were analyzed using EPA Method 1680, and analyses were completed within two hours of collection. The utility's sampling overlapped the project team's visit. The sample collected on July 23, 2007, contained 179 MPN/g dry wt of fecal coliforms. For comparison purposes, the range and variation in the seven samples collected as part of this project were compared to the last four yearly series of samples in Table 4-7. All *Salmonella* testing by the utility in 2004, 2005 and 2007 resulted in <3 MPN/4 g dry wt. The *Salmonella* values for 2006 are presented in Table 4-7. Based on these data, collection and analysis of multiple samples from multiple days captures more variability than analysis of multiple samples on a single day.

Table 4-7. Comparison of Microbial Analyses of MW Utility 3 Samples.

Sample	WERF fecal coliform (MPN/g dry wt)	Utility (7/17/07 - 8/1/07)	Utility (10/26/06 - 11/14/06)	Utility (1/1/05 - 12/5/05)	Utility (7/1/040- 07/20/04)	*Salmonella* 2006 (MPN/4 g dry wt)
1	16	5	41	210	<1	>246
2	39	179	131	8	<1	2.7
3	25	179	281	72	1	<3
4	60	481	7	8	1	6.3
5	17	<1	7	90	13	3.2
6	34	<1	68	1600	<1	<3
7	29	1	37	220	8	<3
Geo mean	32	16	41	73	2	<3.4
Coeff. Var.	36%	862%	175%	506%	179%	1761%

4.1.6 NW Utility 1 – Class B Cake

The influent wastewater for this facility consists of approximately 73% residential, 19% commercial, and 6% industrial inputs. This facility produces their biosolids product by anaerobically digesting gravity-thickened solids that are a blend of primary sludge and WAS. The solids are fed into a blend tank and gravity belt thickened. It is then fed to an anaerobic digester where the target tank temperature is 37°C (98°F) with an average solids residence time (SRT) of 31.7 days. At the end of the digestion process, the solids are pumped to high solids centrifuges for dewatering. The plant maintains four dewatering trains and may run any combination of one to four centrifuges at any given time depending on solids production.

NW Utility 1 produced approximately 12,844 dry tons of biosolids in 2006 that were land applied to fertilize agricultural crops and used to make compost. Based on this production rate, the facility is required to sample once per 60 days, but collects samples monthly. Biosolids samples are collected from specially designed access shafts to the screw conveyors that transport dewatered cake to biosolids haul trucks. Samples are collected every two hours over a 24-hour period and composited for non-biological analyses. In addition, on a quarterly schedule, during the final monthly grab sampling, five sterile sample cups provided to the utility by their analytical laboratory are filled for bacterial and viral analyses and total solids measurement. These five containers are placed in a ziplock bag, placed on ice, and delivered to the analytical laboratory so that bacterial analyses can be completed within 6 hours of collection. The analytical laboratory uses a modification to Method 1680, the LTB/EC multiple tube fermentation method. To prepare the biosolids suspension for inoculation, 50 g of biosolids are mixed with 445mL diluent and 5mL 10% Tween 20 solutions. The material is combined in a sterile pint-sized paint can and mixed on a Red Devil paint can mixer for five minutes. The slurry is then used to inoculate a series of LTB tubes and confirmed in EC tubes as described in Method 1680.

For this project, seven 1L samples were collected on August 7, 2007. Sterile, wide-mouth, plastic bottles were filled to 0.5 to 1 inch of the rim using a dedicated steel sample scoop to reach into the screw conveyer. The sample bottles were placed on ice immediately. The sample for coordinated analysis with the analytical laboratory was produced by compositing equal amounts of samples 5, 6, and 7. The project team samples were analyzed using EPA Method 1680 and analyses were completed within four hours of collection. The analytical laboratory analyzed their sample within three hours of collection using their slurry preparation method. For comparison purposes, the range and variation in the seven samples collected as part of this project were compared to the range and variation reported for samples collected by the utility in 2006. The geometric mean value of sampling conducted in 2005 is also presented. The data are summarized in Table 4-8. Monthly sampling and analysis for biosolids microbial content conducted by the utility captures a similar magnitude of variability as that associated with sampling and analysis of samples from the different centrifuge dewatering trains on a single day. All samples analyzed by the utility were well below the fecal coliform limits for Class B in 2006.

Table 4-8. Comparison of Fecal Coliform Analyses of NW Utility 1 Samples.

Sample	WERF (MPN/g dry wt)	Utility (split sample)		Utility (2006)	Utility (2005)
1	48,300		Min	18,000	--
2	29,400		Max	560,000	--
3	102,000		Median	58,000	--
4	62,200		# Detects	12/12	12/12
5	61,900				
6	50,000				
7	17,900	29,600			
Geo mean	46,900			68,000	39,000
Coeff. Var.	41%				

4.1.7 SE Utility 3 – Class A Compost

This facility, a biosolids processor/vendor, produces Class A compost using sludge and/or biosolids – depending on supplier deliveries – primarily as a nitrogen source in a proprietary mix. The aerated static pile process is used to stabilize Class A and Class B biosolids from municipalities and industries within approximately a 100-mile radius. Each load of feed solids, sludge, or biosolids is weighed and mixed with amendments including wood chips from pallets, yard waste and wood shavings. The piles are 50' long and aerated under cover for 14-18 days using a computerized monitoring process to assure that VAR and PFRP times and temperatures are achieved. The compost is then size-separated using a trommel with a screen opening size of approximately 5/8 inch. Material passing through the screen is placed outside in windrows that are turned daily to cure for 30-45 days depending on the end use, while retained material is returned to the head of the plant and mixed into subsequent batches. Compost may be used for agricultural purposes immediately after the active composting phase. For landscape end uses, the compost is cured, as described above, and may be further mixed with other materials such as soil and bark fines.

SE Utility 3 produced approximately 60,000 tons of Class A compost in 2006. This facility samples final, post-cure compost for nutrients and fecal coliforms on a monthly basis. Composite samples are produced by collecting subsamples from six points in the center of the pile and combining them in two one-gallon plastic bags. The samples are then shipped without ice via UPS for analysis at a contract laboratory (the facility used to ship samples on ice but determined that it didn't make a difference, so stopped).

For this project, seven 1L cured pile samples were collected on June 6, 2007. Sterile, wide-mouth, plastic bottles were filled to the rim using an autoclaved, flat-bladed trowel. The finished compost batches were placed in three-sided concrete bays. Samples were collected from the face of the facility's one completed cured pile, at approximately mid-height. The surface 6 inches of the pile were scraped away using a flat blade trowel. Four holes were dug with a hand auger to a depth of approximately 6-8 inches, approximately evenly spaced across the pile face; the odd-numbered samples were taken at this depth. After the first (odd-numbered) sample for a hole was taken, even-numbered samples were augered below the first one, approximately another 6 to 8 inches into the pile. Two samples were collected per hole, representing mid-depth and deeper samples with the exception of the fourth hole. For the fourth hole, a single sample – sample #7 – was collected by combining material from both depths of the hole. The sample bottles were placed on ice immediately. The samples were analyzed using EPA Method 1680, and analyses were completed within 4.5 hours of collection. Additionally, a composite sample was produced in the laboratory by filling a 150mL specimen cup from each of the seven sample bottles and mixing them on a sterilized foil surface. For comparison purposes, the range and variation in the seven samples and composite analyzed as part of this project were compared to the range and variation reported for the last four series of samples collected by the contract facility. The data are summarized in Table 4-9.

The facility typically samples cured, composted material immediately after it has been screened and placed in the storage bay awaiting transport. The samples collected on June 6, 2007 were from a pile that had been sitting exposed to the elements and wildlife for some time. The results were highly variable with fecal coliform densities spanning almost two orders of magnitude. It should be noted from Table 4-9 that indicator content of a compost pile can vary significantly spatially and with depth in the pile. The data presented in Table 4-9 also confirm that analysis of multiple, individual grab samples of compost provides greater insight into the variability of the material than single samples or composites.

Table 4-9. Comparison of Fecal Coliform Analyses of SE Utility 3 Samples.

Sample	WERF (MPN/g dry wt)	Jan 07	Feb 07	March 07	April 07
1	21,100	2 MPN/g	130 MPN/g	8 MPN/g	300 MPN/g
2	19,200				
3	32,800				
4	6,360				
5	33,600				
6	183,000				
7	67,600				
Composite	66,300				
Geo mean	32,400				28
Coeff. Var.	121%				374%

4.1.8 SE Utility 4 – Class A Compost

This facility produces a composted biosolids product using an in-vessel woodchip system. The "charges" are typically added to each bay daily. These charges are produced by precisely mixing aerobic WAS that has been dewatered to approximated 18% dry solids, with wood chips and recycled compost as a source of seed organisms. The mix targets a finished C:N ratio between 20:1 and 40:1 and a finished compost solids content of approximately 65%. The in-vessel bays are 235' long and turned daily using a computerized process to assure that VAR (55-70ºC (131-158ºF) for approximately 11 days) and PFRP (45-60ºC (113-140ºF) for approximately 32 days) times and temperatures are achieved over a 30-day period. Thermocouples are located every 50' and temperatures are confirmed daily using calibrated probes. The compost is placed under cover to cure for 40 days and is then size-separated to <3/8-inch and <3/4-inch fractions using a trommel screen. The separated, oversized materials are re-chipped and returned to the head of the process. Each batch of cured, size-separated finished compost is stored separately on an open concrete apron. Each batch of final compost is sampled off the concrete apron and tested for *Salmonella* content to meet Class A analysis requirements. The compost is predominantly used for landscape purposes.

SE Utility 4 produced approximately 4,271 dry tons of compost in 2006. As described above, each batch of compost is sampled and tested for *Salmonella* density. Composite samples are produced by collecting discrete samples from 25 points mid-height of the pile around its perimeter. Using an autoclaved large metal spoon, five scoops from each location are placed into a five-gallon bucket lined with a sterile plastic bag. Once all of the sample material is added to the bucket, it is closed and shaken and stirred to mix the contents. A representative sample is taken from this mixture and placed in a sterile container. The composite sample is then shipped in a cooler with ice packs for analysis at a contract laboratory.

For this project, seven 1L samples were collected on June 11, 2007. Sterile, wide-mouth, plastic bottles were filled to the rim using an autoclaved, flat bladed trowel. Samples were collected from the face of the pile approximately mid-height. The surface 6 inches of the pile were scraped away using a flat blade trowel. Seven holes were dug with a hand auger to a depth of approximately 6-8 inches, approximately evenly spaced across the pile face. The first two samples were much wetter than the other samples, owing to being located on the north side of the pile and having been subjected to 1.5 inches of rain approximately four days prior to sampling. The sample bottles were placed on ice immediately and analyzed using EPA Method 1680 within 4 hours of collection. Additionally, a composite sample was produced in the laboratory by filling a 150mL specimen cup from each of the seven sample bottles and mixing them on a sterilized foil surface. At the time of sampling, the facility was undergoing an intensive microbial study and had been analyzing their compost for fecal coliforms and *Salmonella* since February 2007. For comparison purposes, the range and variation in the seven samples and composite analyzed as part of this project were compared to the range and variation reported for the last four samples collected by the contract facility. In addition, the compost pile sampled by the project team had also been sampled and analyzed by SE Utility 4 on April 17, 2007. This monitoring data is summarized in Table 4-10.

The methodology by which the utility produces their composite sample appears to be achieving a sample representative of the biosolids batch. It is important to note that samples of batch C-1 were collected by the utility and project team almost two months apart. The resultant fecal coliform density was 75 MPN/g dry wt for the utility composite, and the geometric mean of the individual grab samples analyzed by the project team was approximately 62 MPN/g dry wt. Using this comparison, the scheme used by the utility to produce their composite sample appears to be achieving a representative sample. However, the WERF project team composite value was 1,430 MPN/g dry wt. This indicates that using composite samples or reporting only the geometric mean might mask the presence of non-503 compliant hotspots. The pile of batch C-1 had been stored outside on a concrete pad for a number of months. A rain event wetted the pile prior to the project team sampling event and bird activity was noted around the pile storage area. The "wet" areas of the pile exhibited the highest FC results, indicating possible regrowth. Additionally bird droppings have been suspected by the utility to reintroduce *Salmonella* on storage. Once again, analysis of multiple individual samples provides greater insight into the variability within the pile.

Table 4-10. Comparison of Bacteriological Analyses of SE Utility 4 Samples.

Sample	WERF C-1 6-11-07 fecal coliform (MPN/g dry wt)	Utility C-1 4-17-07 (MPN/g dry wt)	Utility C-7 3-26-07 (MPN/g dry wt)	Utility C-5, C-6 3-26-07 (MPN/g dry wt)	Utility C-6 3-12-07 (MPN/g dry wt)
1	2,390	--	--	--	--
2	<4.2	--	--	--	--
3	<2.5	--	--	--	--
4	<2.5	--	--	--	--
5	3,160	--	--	--	--
6	3.4	--	--	--	--
7	5,380	--	--	--	--
Composite- FC	1,430	75	<0.30	13	36
Composite-*Salmonella*		<0.30	<0.30	<0.32	<0.36
Geo mean*	62.3				
Coeff. Var.	5362%				

*the detection limit was used for calculation of the geometric mean and may be a conservative overestimate.

4.1.9 NE Utility 1 – Class A Compost

This facility serves a community that consists of a mixture of residential and industrial service connections. While the design capacity of the plant is 2.2 mgd, typical flow is about 1.3 mgd. This facility produces composted biosolids using an aerated static pile woodchip system. Anaerobically digested sludge (a mix of primary solids and WAS) is dewatered and mixed with commercial woodchips. The piles undergo active composting for 21 days. A continuous thermal probe monitors pile temperature throughout those 21 days. The pile must be maintained at >45°C (113°F) for 14 days and >55°C (131°F) for three days to meet pathogen destruction and vector attraction reduction time and temperature requirements. The material is then allowed to cure for 52 days. Once curing is complete, the compost is screened using a trommel device to separate oversized materials and recover woodchips that are returned to the head of the process. Each batch of finished, cured and size-separated compost is stored separately on an open-sided, but covered concrete apron. The compost is predominantly used for landscape purposes.

Once a quarter, the compost pile is sampled and tested for *Salmonella* content to meet Class A microbial monitoring requirements once every 90 days. Two 1L containers are filled with screened compost from the most recent material added to the storage pile.

For this project, seven 1L samples were collected on July 9, 2007. Sterile, wide-mouth, plastic bottles were filled to the rim using an autoclaved, flat-bladed trowel. Samples were collected from the face of the pile approximately mid-height. The top six inches of the pile were scraped away. Seven holes were approximately evenly spaced around the perimeter of the pile. The samples were analyzed using EPA Method 1680, and analyses were completed within four hours of collection. Additionally, a composite sample was produced in the laboratory by filling a 150 mL specimen cup from each of the seven sample bottles and mixing them on a sterilized foil surface. The data are summarized in Table 4-11. Once again, it can be seen that the variability in fecal coliform content of composted material is highly variable. In this study, two of seven samples met Class A levels for fecal coliforms, while for five samples, the composite, and geometric mean were above Class A limits. However, for testing for *Salmonella* demonstrated the material to always be below the Class A limits.

Table 4-11. Comparison of Microbial Analyses of NE Utility 1 Samples.

Sample	WERF (FC/g dry wt)	Utility (January 07) *Salmonella* (MPN/4 g dry wt)	Utility (April 07) *Salmonella* (MPN/4 g dry wt)	Utility (July 07) *Salmonella* (MPN/4 g dry wt)
1	261,000	<3	<3	<3
2	161,000			
3	37			
4	139			
5	1,390			
6	9,870			
7	23,800			
Composite	14,300			
Geo mean*	4,920			
Coeff. Var.	758%			

*the detection limit was used for calculation of the geometric mean and may be a conservative over estimate

4.2 Brief Study Site Results

4.2.1 SW Utility 1 – Class A Liquid

The service area of this facility includes a combination of residential and commercial/industrial inputs. This facility produces their biosolids product by processing secondary solids. WAS is thickened and fed into one of three anaerobic digesters (total combined volume 2 million gallons) where the solids residence time (SRT) is about 25 days. The solids are transferred to one of six (totaling 25 acres) facultative lagoons, where the material may be stored for 1 to 4 years, depending on the utilization of storage capacity. At the end of the storage time, the depth and quality of solids at the bottom of the lagoon are tested. The material is then pumped onto a 50-acre dedicated land disposal (DLD) site. Approximately 80,000-120,000 gallons of WAS is processed each day.

SW Utility 1 staff was interviewed as to their sampling protocol. This facility is required to collect and analyze a composite sample once per quarter. However, sampling is typically conducted prior to pumping of the lagoon to the DLD. For 2007, this resulted in bi-monthly sampling.

The composite sample for SW Utility 1 is produced by combining 20 grab samples into a carboy lined with a sterile teflon bag. The material is well mixed and split for a variety of analyses. Individual samples are collected from a boat using a plexiglas Kemmerer sampler. Fifteen sampling points are chosen around the perimeter of the lagoon, and the five remaining samples are collected along a transect across the middle of the lagoon. Once the sampler is deployed and filled, the sampler is emptied into a Teflon-bag lined bucket, mixed, and used to fill a sterile, 16oz, plastic sample container. The sample containers are placed on ice in a cooler, and the next lagoon location is sampled. The laboratory is located within 10 minutes from the lagoon site, and the composite is produced upon returning to the laboratory. The composite intended for microbial analysis is placed on ice and hand delivered to the contract laboratory. This allows for analyses to be conducted within six hours of collection. The contract laboratory analyzes for fecal coliforms using Method 1681, *Salmonella* using Method 1682, enteric viruses using ASTM D 4994-89, and helminth ova using EPA 600/1-87/014.

Results from the most recently completed year of sampling rounds are summarized in Table 4-12 below. The levels of fecal coliforms varied by less than an order of magnitude between sample dates, indicating the effectiveness of this utility's lagoon treatment system.

Table 4-12. Microbial Analyses of SW Utility 1 Samples.

Sample	Utility (4/19/2007)	Utility (6/20/2007)
Fecal coliform (MPN/g total solids)	450	96.3
Salmonella (MPN/4 g total solids)	<0.7	<0.5
Enteric virus (pfu/4 g total solids)	<1	<1
Helminth ova (ova/4 g total solids)	<1	<1
Total solids	3.72%	5.12%

4.2.2 NW Utility 2 – Class B Cake

The service area of this facility consists of 77% residential, 15% commercial and 8% industrial. This facility is a conventional activated sludge wastewater treatment plant with a rated wet weather capacity of 115 million gallons per day (mgd). Solids handling facilities consist of six dissolved air flotation tanks (DAFTs) for sludge thickening, four mesophilic anaerobic digesters for sludge VAR and PSRP and three high-solids centrifuges for dewatering. Flow through a one to two hour blending tank after the DAFTs helps to generate a homogenous product to feed to the digesters. Similarly, a two to five day blending tank after the digesters helps generate a homogenous product to feed the dewatering process. The digester operating temperature is 37°C (99°F); solids retention time averaged 24.9 days in 2006.

NW Utility 2 produced approximately 13,271 dry tons of biosolids in 2006 that was land applied to fertilize agricultural crops and forests, and to make compost. Based on this production rate, the facility is required to sample once per 60 days, but actually samples monthly. Biosolids samples are collected from specially designed access shafts to the screw conveyors that transport dewatered cake to holding tanks. Samples are collected every two hours over a 24-hour period and composited for non-biological analyses. In addition, on a quarterly schedule, during the final monthly grab sampling, five sterile sample cups provided to the utility by their analytical laboratory are filled for bacterial and viral analyses and total solids measurement. The five containers are placed in a ziplock bag, placed on ice, and delivered to the analytical laboratory so that bacterial analyses can be completed within 6 hours of collection. The analytical laboratory uses a modification to Method 1680, the LTB/EC multiple tube fermentation method. To prepare the biosolids suspension for inoculation, 50 g of biosolids are mixed with 445 mL diluent and 5 mL 10% Tween 20 solutions. The material is combined in a sterile, pint sized paint can and mixed on a Red Devil paint-can mixer for five minutes. The slurry is then used to inoculate a series of LTB tubes and confirmed in EC tubes as described in Method 1680.

While this site was intended to participate as a "brief study site," opportunity allowed for the project team to complete limited sample analysis. For this project, three 1L samples were collected on August 7, 2007. Sterile, wide-mouth, plastic bottles were filled to 0.5 to 1 inch of the rim using a dedicated, steel sample scoop to reach into the screw conveyer. The sample bottles were placed on ice immediately. Two centrifuge dewatering trains were in operation at the time of sampling. Samples 1 and 2 were collected from one dewatering train, and sample 3 and the utility's analytical lab sample were collected from the second dewatering train. The project team samples were analyzed using EPA Method 1680, and analyses were completed within 4 hours of collection. The analytical lab analyzed their sample within two hours of collection using their slurry preparation method. For comparison purposes, the range and variation in the three samples collected as part of this project were compared to the range and variation reported for samples collected by the utility in 2006. The data are summarized in Table 4-13. The geometric mean value of the results of sampling conducted in 2005 is also presented. Monthly sampling and analysis for biosolids microbial content conducted by the utility captures greater variability than that captured by sampling and analysis of samples from the different centrifuge dewatering trains on a single day. Sampling multiple batches provides greater insight into a utility's process variability. All samples were well below the fecal coliform limits for Class B of 2 million MPN per gram dry wt.

Table 4-13. Comparison of Fecal Coliform Analyses of NW Utility 2 Samples.

Sample	WERF (MPN/g dry wt)	Utility (split sample)		Utility (2006)	Utility (2005)
1	154,000	81,000	Min	22,000	--
2	51,900		Max	640,000	--
3	115,000		Median	61,000	--
			# Detects	12/12	12/12
Geo mean	97,200			74,200	28,000
Coeff. Var.	38%				

4.3 Key Phase III Findings

The field visits and interviews produced several observations and key findings.

- The importance of tank mixing was recognized and utilized at all three of the facilities producing a liquid product.
- The importance of wasting liquid before sample collection was recognized at the facilities that collected liquid samples with pumps, taps, or side spigot.
- Capturing biosolids variability, and possibly obtaining a more "representative" sample, was achieved at liquid facilities by analyzing samples collected on multiple days.

- Sampling biosolids immediately after completion of the active process phase may be representative of indicator and pathogen densities in a biosolids batch; however, sampling over time, as most utilities do, captures greater process variability.
- Sample collection and analysis of stored material (both compost and one liquid) results in higher densities and more highly variable indicator values.
- The importance of using sterilized containers and sampling devices was recognized by facility personnel at all facilities visited.
- The importance of short transport time and icing of samples was recognized by all facility personnel; however, detailed guidance on icing techniques would be welcomed.
- The facility personnel were receptive to the possibility of using "field sterilization" techniques to reduce the numbers of sampling devices and cross contamination between individual samples collected to produce composites.

CHAPTER 5.0

COMMUNICATIONS

Many in the wastewater industry recognize the challenges associated with public confidence and understanding about treatment processes and the safe production of by-products from these processes for land application. Industry representatives also realize the importance of communications with regulatory officials, community leaders, and within the industry and individual operations. Because this project's focus is on sampling biosolids for microbial analysis, an attempt was made to create a set of tools primarily addressing this particular issue. The approach taken in this work was intended to result in products that fit within a broader context of communicating about biosolids and that serve as examples for which other specific issues could be similarly addressed.

The following information provides a summary of the approach and evolution of the communications component of this project. Several sample communication pieces and/or tools resulted from this effort and are included in Appendix B. Each of these items were informed by other aspects of this research project, including the survey results, literature reviews, in-field sampling in Phases II and III. They were subject to beta-testing with participating utilities.

5.1 The Communications Approach

A well-derived communication strategy is based on identification of a set of overarching issues surrounding a given topic or sector of activity, each important to effectively communicate. It also requires knowledge of the diverse stakeholders that will have some level of interest in the issues as a result of real or perceived impacts they may face. Because wastewater treatment systems are a service with public benefit, there are a wide range of issues that may be of interest to certain audiences. These treatment systems are very diverse in terms of functional design, operations and management, thus creating more need for effective communications among stakeholder groups. The by-products of wastewater treatment processes may be considered a renewable resource and must be safely and efficiently managed by plant operations, particularly the biosolids materials resulting from various solids separation and treatment processes. Communications that result in confidence about the safety of these materials for beneficial use is of critical importance.

Communication about the focal topic of this WERF project, *Biosolids Sampling Practices for Microbial Analysis*, could be considered within the broader context of wastewater treatment and resource reclamation practices, or as a stand-alone issue for which a specific communications strategy is defined. Regardless, it is expected that diverse stakeholders would benefit from effective communications. Provided below are a few examples of potential stakeholders that may be interested in issues pertinent to the overall wastewater operation, or concerned with sampling only. Those shown in italics are the groups where an overlap in communications efforts would undoubtedly occur.

Wastewater Management Stakeholders

- *Regulatory agencies*
- *Local governments*
- *Public health professionals*
- *Technology vendors*
- Engineers
- *Utility operators/ managers*
- Utility professional groups/*organizations*
- *Researchers*
- *Users* of biosolids products
- Contract management entities

Biosolids Sampling Practices Stakeholders

- *Regulatory agencies, public health officials*
- *Vendors*
- Professional waste & water associations
- Laboratories
- *Researchers*
- *Operations managers*
- Technicians and samplers
- *Contract management entities*
- Product end-users

Within the communications framework and strategy for this particular project, attention was focused on the following specific stakeholder groups for creating product and interfacing examples:

- Technicians and samplers
- Plant operators/managers
- WERF as a professional organization
- Regulatory compliance & public health officials

Communication tools can take on many forms and may change as the strengths and weaknesses of each are evaluated once a strategy is implemented. Regardless, the tools for this project's strategy require that technical, policy, regulatory, science and research issues be addressed. These tools should help improve awareness at the community, agency, practitioner and end-user level.

Effective implementation of the communication strategy requires understanding the use and benefits of the different tools. Considerations include:

- frequency of delivery and availability of content
- frequency of updates to time-sensitive content
- outreach mechanisms
- message testing
- trust building
- feedback loops
- a suite of tools

- fact sheets
- web links
- tri-folds for specific audiences
- community bulletins
- posters
- kiosks and poster boards
- training videos and design manuals
- workshops
- blogs
- pocket guides

Some tools can work in tandem with other tools, such as fact sheets with poster boards, community bulletins with kiosks, pocket guides with posters, etc.

5.2 Purpose of the Communications Strategy

A well-conceived strategy facilitates a better understanding of needs, concerns and opportunities facing different stakeholders. Such a strategy clearly articulates interfacing issues of biosolids science and quality assurance techniques; compliance, health and safety; and operations efficiency. Effective implementation increases stakeholder awareness and fosters improvements in biosolids management and quality control.

Each product derived for the specified stakeholder audience(s) focuses on the primary issues relevant to that group. They also inform other stakeholders about key concerns. Each tool or piece is developed with a particular intent.

- The *technical* tool concisely explains accepted or newly proposed biosolids *sampling approaches* for microbial analysis, and the rationale (*science, safety*) behind each approach.
- The *compliance* and public health related piece increases awareness about existing sampling and analysis requirements, the rationale (*health, safety, quality)* for those requirements, and the areas where changes may be beneficial.
- The *operations and management* communications inform staff, contractors and other audiences of how managers maintain compliance and quality controls.

5.3 Other Potential Communication Efforts

There are several products that could be utilized as part of a communications effort. Sources of information resulting from this project may or may not enhance broader communications goals. These products include a research digest, the executive summary, this final report, a sampler guidance booklet, peer-reviewed manuscripts, and conference presentations.

Primary elements of the communications framework should be concise, stakeholder-specific when referenced and/or utilized, understandable to target audiences, and clearly seen as integral components for improving knowledge.

5.4 Communications Beta-test Results Summary

The goals of the communications beta-test were to assess the potential usefulness of the proposed communications strategy and, more specifically, some products that can be available to stakeholders. The inquiry determined what information or issues need improved communication, and it identified key audiences.

Beta-test participants were utilities involved in Phase III onsite sampling with Dr. Long, or had previously indicated an interest in participation. A list of the beta-test recipients, the questionnaire, and explanation of communications tools are included in Appendix C. Participants received the content, without special formatting, for the Sampler Technical Guidance, Operations and Manager News & Notes, Internal to Staff, Community Bulletin, and Compliance and Health Bulletin. Formatted versions of News & Notes, the Compliance and Health Technical Bulletin, and the Sampling Poster were also provided for comment. Many of the responses to the questionnaire were very informative and articulate; therefore, some quoted responses are provided in an "example of answers" summary included in Appendix C.

5.4.1 General Comments and Feedback of Note

The poster was referenced by most respondents as a very useful and practical tool. Some recipients commented they could see creating a modified version of the poster as a tri-fold for distribution. Most agreed the content was concise and to the point, the visuals were a positive contribution to content, and that in some instances emphasis within any one of the three sections could be modified as necessary. For example, several respondents felt the first section about process was as important for samplers to understand as the rest of the poster. A couple of respondents discussed the critical nature of including the shipping instructions, manifest requirements, etc., particularly in their state.

Most of the respondents agreed the guidance for sampling, specifically for microbial analysis, is not well communicated, inconsistent and in some cases deemed impractical. Several respondents commented they felt there should be a set of clear specifications for Class A as well as Class B biosolids materials. A number of respondents land apply their biosolids, and although they felt the overall content of the communications tools were potentially of use and that, in general, efforts to improve communications are positive, they were typically sending their materials to the same farmers and/or had acquired their own land for application and therefore did not see as big of a communications need. Some of these same individuals indicated that farmers typically want to know more about the agronomic benefits than any other information, with several indicating they actually conduct soil sampling for their customers and help determine proper application rates. Specific to sampling practices, the beta-test revealed that truck drivers/haulers often are the ones who end up taking biosolids samples, either while loading the unit or prior to distribution onto the land. Therefore, a need was identified to improve communications with those stakeholders to increase their awareness of the purpose and need for proper protocol.

Several utilities are in the process of developing an environmental management system (EMS) program and, as part of that process, require communication tools and strategies. They felt that some of the tools would fit nicely into their developing program. Similarly, other utilities felt they could use some of the tools in tandem with their standard operating procedures (SOP) for the utility.

Many of the respondents felt the relationship between the wastewater treatment and biosolids handling processes, and the sampling required to demonstrate end-of-process quality, needed to be better understood and communicated to different audiences. Most indicated the main reason for sampling, and ultimately any effort to communicate outside of their operation, relates to compliance with regulations and further demonstrating they are in compliance.

Following the "Sampling Poster," the Operators and Managers "News & Notes" was the most popular piece. Respondents indicated the proper use of text boxes, bullets, and pictures or visualization tools, is useful.

5.4.2 Comments Specific to Content

When asked if the *general* content of the various communication product examples was potentially useful, most respondents indicated yes. A few respondents felt information to deal with the public is more important than for staff, with some indicating that the pieces for the staff and the poster were good. However, for other audiences, content appeared to express general assumptions. The operations that employ an EMS indicated the content was helpful and could fit in or complement their current efforts.

Nearly all respondents agreed that the most important *specific* communications content relates to public perception about the *safety* of biosolids, followed by the benefits of its use. Several liked the feel of the documents – making use of "we," "you," etc. – to make them more personal and more like an effort to reach out and communicate. Others mentioned they were very focused on data that they share most with local officials and in-house operators, typically putting the data in graphic form. The timely availability of information regarding procedures must be made available to staff as part of compliance-based communications. Specific to sampling, the content addressing what is being done with samples, why they are taken, and why it is important to apply specific methods was deemed helpful; lab personnel and technical experts know these things, but haulers and onsite samplers often don't. A few respondents felt it would be helpful to be more pointed about how potential sampling mistakes can have an impact on the environment.

When asked if the use of more visual content would be beneficial in communicating with their audiences, overall respondents agreed the use of visuals is important, helps to break up boring text and, for some, is the best way to help with comprehension. Many recognize that some technical or scientific verbiage needs visuals along with it, and felt actual pictures showing what is going on helps with understanding (i.e., spraying, taking a sample, analyzing a sample, using materials in different ways, shipping, etc).

5.4.3 Formats

As previously mentioned, the poster format was well received, not only because it was useful, but it also efficiently condensed information. The News & Notes received mixed responses, with some seeing the format as a good way to highlight certain content and effectively add visuals and pictures. The technical bulletin, though acceptable content-wise, contained too much text to be broadly effective in the view of some.

Others commented that the News & Notes could be used for several external audiences and across departments within the operation. Bulleted formats are fine for use at the plant, but the more attractive, visually oriented materials are needed to capture the attention of outside audiences. Some felt the formats and content were very informative and reflected a proactive attitude, which is good. Others indicated they like the template approach, and that having the content and templates electronically available would be helpful. A few respondents indicated they did not think these forms of communication would be that useful because they likely would not progress past two or three editions.

When asked about products being delivered as different forms of electronic files, most indicated they did not think this would limit their ability to use. Some of the larger utilities have staff trained in production of materials and have broad capabilities. Many indicated they have access to multi-media services or print-shops that can handle most anything, but cost could be an issue. The question might not have been fully understood with regard to using different production software as opposed to "electronic file" capabilities, as all respondents indicated they are good at or getting better with electronic file handling.

In responding to the question about simple or more visually intriguing products, the majority of respondents agreed that in certain instances, particularly for outside audiences and those that are not operations/technically oriented, the use of visuals and more attractive pieces is necessary. This is, in part, to get positive attention and awareness drawn to the communication, which can mitigate the highly variable levels of education and comprehension skills stakeholders may have. Some participants thought staying as simple as possible was best, particularly for internal audiences and more personalized communications.

When asked how they currently distribute information, respondents indicated the use of different methods, with many indicating they utilize the web (e-mail, website) as a primary tool. Some indicated they use a customer newsletter, a "city-page" in a local newspaper, utility water bill inserts, and sometimes verbal communications.

5.4.4 Notable Communication Needs

Most of the respondents indicated they believe the public remains misinformed and lacking in information about biosolids in general. Whether the issue is about sampling or processes, odor control, or economic benefits, the public and some officials need assistance understanding what is really being done at these facilities and why. Technical information needs to be conveyed, but in ways easily understood.

Specific to the beta-test and the focus on sampling, the answers to the questionnaire were variable. Overall, many respondents felt the frequency of sampling, the reasons for having QA/QC and for following proper protocol, and the implications that different processes and tests have on how to sample and ship biosolids materials are most important. Responses again varied as far as what information concerning microbial analysis sampling needs to be better communicated. However, many indicated the frequency of sample collection and treatment processes the utility has in place, and their relevance to biosolids sampling needs, may be most important to understand.

Others mentioned a need to communicate sampling requirements in relationship to storage of biosolids, and being able to help samplers understand the impacts that changes in laboratory analysis procedures can have on sampling activities. Related to the latter, some observed a need for improved communications between laboratory staff and onsite samplers.

Others noted the importance of communications as a means to insure staff is well trained, and that there is continuity (knowledge transfer) as staff turnover occurs.

Several respondents indicated a need exists to communicate with designers and engineers in an effort to avoid poor process designs that ultimately can have an impact on sampling procedures. A few participants commented that regulations seem to cover a narrow range of indicators/pathogens; that regulatory review processes, though necessary, don't necessarily validate current practices, and that potentially useful ecotoxicology studies are few and may be important for public confidence.

Other audiences relevant for targeted communications that recipients mentioned included boards and commissions, policy-makers, agricultural users, special-interest groups, regulatory agencies, the land application community, city council members, and children. As far as audiences within the operation, several respondents mentioned levels of management, samplers and monitoring personnel, in addition to the haulers and applicators.

When asked whether it would be beneficial to recipients if "your industry" implemented a communications strategy and provided similar types of tools or products to stakeholders for your operation to distribute, with the exception of a few respondents, nearly all participants indicated that having an industry-wide strategy and access to useful, consistent and informative communication tools would be beneficial. Respondents generally commented that improving communication is a plus and is necessary.

Some responses suggested better standardization of sampling protocols and creation of associated communication tools for standardized procedures. Some felt there are certain instances where this would be beneficial, but it would be issue-dependent and they were not sure that sampling practices are that critical in terms of communications, whereas overall safety of biosolids and explanation of processes in relationship to compliance was more important.

5.4.5 Final Note

The beta-test proved to be a good source of insights concerning the issues that are believed to be in need of better communication, the content and formats worthy of more focus, and the recognition of variability in practices and outreach efforts across the industry. In creating the final version of the tools for this project, feedback from the beta-test was integrated into the content and choice of format alternatives. In addition to this report summary, there are examples of specific responses to the questionnaire included in Appendix C. The communications component of this project proved to be informative and appears to have added value to the overall results.

CHAPTER 6.0

RECOMMENDATIONS AND FUTURE RESEARCH NEEDS

6.1 Recommendations

The following bullets highlight the significant findings from the information gathering and experimental work of this project.

- Storage tanks and digesters from which liquid biosolids are being sampled should be well mixed before the sample is collected.
- Precise measurement of purging 10 pipe derived volumes is not necessary prior to liquid sample collection provided that all dried or caked material on pipe openings are removed before sample collection. Wasting of initial pipe contents is still recommended.
- Analysis of multiple samples of biosolids (liquid, cake and compost) appears to be more capable of identifying variability in microbial content than production and analysis of composite samples (see Table 3-7 for optimum numbers).
- Capturing biosolids variability, and possibly obtaining a more "representative" sample, can be achieved by collecting and analyzing individual samples collected on multiple days.
- One liter samples, as specified in EPA Methods 1680, 1681, and 1682 for biosolids analysis, appear adequate in size.
- Overnight shipping of samples on ice (up to 24 hours of storage before analysis) does not appear to significantly alter the numbers of fecal coliforms enumerated from liquid or cake biosolids using Methods 1680 and 1681.
- Field sterilization methods appear adequate to prevent cross contamination of samples as long as the guidance is clear and adhered to strictly.
- Multiple 30 gram replicates or larger proportions of biosolids to buffer slurries are required when analyzing for *Salmonella* and viruses in order for detection limits to be below the regulatory levels for biosolids of less than 9% solids content.
- Sampling and microbial analysis of multiple discrete samples of cake material (produced from anaerobic digestion followed by centrifuge dewatering) should be conducted after storage but prior to distribution.

The review of sampling schedules and practices at the nine field sites and two interview sites in conjunction with experimental work results in the following recommended practices: 1) utilities should sample biosolids as frequently as their budgets and labor constraints allow, but at least once from each batch of biosolids; 2) multiple grab samples rather than composites should be analyzed; 3) one liter samples should be collected and properly cooled during storage while awaiting analysis within holding times specified in the Federal Register.

6.2 Future Research Needs

The project results must be considered in the broader context of biosolids product quality assurance and risk mitigation. While some distinct questions have been answered, the heterogeneity of sludge and biosolids products, the lack of precision of existing analytical methods and the great diversity in biosolids treatment practices, treatment trains and specific technologies as well as the natural variation within each of these areas (e.g., seasonal, inter- and intra-batch, etc.) limits the extent of statistical inference than can be made. Phase III results indicate that storage of biosolids can result in significant changes in indicator organism (fecal coliform) densities through opportunities for regrowth. The implications affecting microbial risk are unclear. Examples from the drinking water engineering literature indicate that pathogens do not regrow under environmental conditions that allow coliforms to regrow. It is untested as to whether this is true for the biosolids environment.

Accordingly, future research should focus on addressing sampling and handling methodologies while considering factors most likely to be influential and those which can be measured or controlled sufficiently to permit the collection of statistically robust data within reasonable research budgets. Additionally, future research and the results thereof must be analyzed and communicated with attention to quantifying risks associated with the statistical limitations. Specific areas which the research team believes should be addressed in future research include:

- validating sample collection and handling techniques for a limited number of commonly used biosolids processing trains;
- investigating the viability of a protocol which incorporates baseline product sampling supplemented with less-intensive, facility-specific periodic sampling for assuring product quality, environmental and health protection and regulatory compliance;
- determining minimum sampling and handling requirements for small or lower budget facilities and those producing biosolids products for end uses with minimal exposure risks; and
- determining whether sampling protocols appropriate for other biosolids criteria (e.g., heavy metals, nutrients, organics) overlap in effectiveness for microbial testing.

CHAPTER 7.0

KEY RESEARCH ACTIVITIES AND FINDINGS

The following summarizes the issues, key research findings, information gaps, and research needs associated with the following sampling needs: preparation, sampling of liquid biosolids, sampling of cake or compost from piles, sampling of cake from conveyors, and handling and shipping.

Preparation

Issues: sterilization and container size and type

Key findings: containers can be purchased pre-sterilized or sterilized using an autoclave; one-liter glass and plastic containers are acceptable and most prevalent in use; stainless steel sampling tools (scoops, trier samplers, thief samplers, augers) may be washed, wrapped and autoclaved before use

Information gaps: field sterilization of sampling tools is allowed under the 503 regulations; however, proper protocols need to be developed and disseminated

Research needs: none identified

Sampling of Liquid Biosolids

Issues: sampling frequency, sampling location, sampling approach, sampling methodology

Key findings: each batch as budgets allow (which may be more frequently than prescribed in the 503s based on production volume); from side taps as liquid is being pumped to storage/trucks or from storage tanks with active mixing; one liter samples into bottles leaving 0.5 inch headspace for gas production

Information gaps: how to produce composite samples for liquids stored for a significant amount of time; experimentally-verified guidance on purging of pipes and side taps

Research needs: is there a difference among results for samples collected using a dipper, bailer, hand pump, or side spigot/tap

Sampling of Cake or Compost from Piles

Issues: sampling frequency, sampling location, sampling approach (grab vs. composite), sampling methodology

Key findings: each batch as it is added to storage piles as budgets allow; for fresh product multiple discrete samples using a random grid; for stored product a composite sample including up to 50 subsamples using a random grid; 150 mL or 4 oz. specimen cups may be used to collect subsamples to form a composite

Information gaps: what is the definition of stored product

Research needs: can storage logisitics (closed vessels, open bays, outdoor piles, etc.) affect sampling needs

Sampling of Cake from Conveyors

Issues: sampling frequency, sampling location, sampling approach, sampling methodology

Key findings: multiple samples of each batch as budgets allow (which may be more frequently than prescribed in the 503s based on production volume); one liter samples across a cross section of the conveyor

Information gaps: does dewatering process affect where in the process should samples be collected with regards to reactivation and regrowth phenomenon

Research needs: same as information gaps

Handling and Shipping

Issues: cooling regimes, holding times

Key findings: a cooler should be on hand at sampling location and contain enough ice or gel packs to surround the sample bottle; samples should be analyzed as soon as possible (within 24 hours for fecal coliforms, six hours for *Salmonella*, two weeks frozen for total culturable virus, four weeks for viable helminth ova)

Information gaps: none identified

Research needs: does the biosolids production process affect sample holding times (*i.e.* stability of microbial population)

APPENDIX A

UTILITY SURVEYS

A-1 Manager's Survey
A-2 Sampler's Survey

A-1 Manager's Survey

UMBS-###m

Survey of Biosolids Sampling Practices

1. What types of biosolids materials (or products) are produced by your utility? (check all that apply)
   - □ Cake (including granular)
   - □ Liquid
   - □ Compost
   - □ Other ____________________

2. Is any part of sludge processing/final biosolids product production prior to beneficial reuse contracted out?
   - □ Yes (specify name of company: _______________)
     - __________% receives final processing in house
     - __________% receives final processing by an outside contractor/vendor
   - □ No

3. Final Disposition of Biosolids
   A. What are the end use(s) of the biosolids that receive final processing by your utility **in house**? (check all that apply)

   - □ Not applicable – all final processing is done by an outside contractor/vendor.
   - □ Residential use
   - □ Commercial use
     - o Landscaping
     - o Agriculture
   - □ Landfill (daily cover and/or disposal)
   - □ Incineration
   - □ Land disposal (non-beneficial use)
   - □ Other ____________________

   B. What is the final disposition of biosolids that receive final processing **by an outside contractor/vendor**? (check all that apply)

   - □ Not applicable – all final processing is done in house by the utility.
   - □ Residential use
   - □ Commercial use
     - o Landscaping
     - o Agriculture
   - □ Landfill (daily cover and/or disposal)
   - □ Incineration
   - □ Land disposal (non-beneficial use)
   - □ Other ____
   - □ Not sure ____

4. Does your utility apply any type of quality assurance and quality control (QA/QC) program to microbial/pathogen sampling for biosolids products?

   - □ Yes, we have a quality assurance and quality control program
   - □ No, we do not have a quality assurance and quality control program. (please skip to question 9).

5. Do you have a **written** quality assurance and quality control plan for microbial/pathogen sampling for biosolids? If so, please list the major program objectives.

   - □ Yes, the objectives are: ____

     ____

   - □ No

6. Please describe your utility's quality assurance and quality control program for sampling procedures for biosolids materials intended for microbial/pathogen analysis.

7. Is your QA/QC program specific to certain types of biosolids materials and/or end use(s) for which the materials produced?

   □ Yes, it is material or end-use specific. Please describe it briefly:

   □ No, it is not material or end-use specific.

8. If you contract sludge processing services with an outside vendor do they require provision of a QA/QC program?

   □ Not applicable – we do not contract out any biosolids processing.
   □ Yes, the vendor requires that we follow a QA/QC program.
   □ No, the vendor does not require us to follow a specific QA/QC program.

9. Does your utility conduct microbial/pathogen analysis of biosolids samples onsite?

   □ Yes, we conduct microbial/pathogen analysis on site.
   □ No, we send our samples off-site for microbial/pathogen analysis.
   □ No, we do not conduct microbial/pathogen analysis.

*The remainder of this survey concerns only one treatment facility managed by your utility. Use questions 10 through 11 to identify which facility. This facility will be referred to as the **study facility**.*

10. Does your utility have a centralized sludge/biosolids processing facility (instead of individual sludge/biosolids processing facilities at each WWTP)?

    □ Yes → Go to question 12 and use this biosolids processing facility as the *study facility*.
    □ No

11. Does your utility have just one wastewater treatment plant?

    □ Yes → Go to question 12 and use this plant as the *study facility*
    □ No → Choose the largest plant that produces biosolids cake, liquid, and/or compost and use this plant as the *study facility*.

12. What is the wastewater treatment plant capacity for the study facility?

_______ MGD; ____________ connections served, __________% flow from industry.

13. What types of biosolids materials are produced at the study facility? (Check all that apply.)
   - □ Cake (including granular)
   - □ Liquid
   - □ Compost
   - □ Other ______________________

14. What class of biosolids product is produced at this plant and what alternative(s) is/are used to meet and determine compliance with pathogen requirements (*e.g.*, per Chapters 4 and 5 of EPA's "White House Document", *Control of Pathogens and Vector Attraction in Sewage Sludge,* EPA/625/R-92/013, revised July 2003)?

   - □ Class A / alternatives used:____

   - □ Class B / alternatives used:____

   - □ Not Classified

15. When was the study facility initially built? (If unsure, give an approximate year.)

16. When was the last time the study facility had a major expansion or renovation, if ever?

   - □ There haven't been any major renovations or expansions to the study facility
   - □ The last major renovation occurred in (year) ________________
   - □ The last major expansion occurred in (year) ________________

17. Wastewater Treatment Plant (Liquid Treatment) Processes at the study facility (check all that apply)

□ Not applicable – the study facility is a centralized sludge/biosolids processing facility.

- □ Screening
- □ Grit/grease removal
- □ Primary settling
- □ Secondary (biological) treatment
- □ Secondary clarification
- □ Tertiary filtration
- □ Chemical precipitation
- □ Powdered activated carbon treatment
- □ Membrane filtration
- □ Ion exchange
- □ Other __________________________
- □ Activated sludge (aerated suspended growth process)
- □ Anaerobic suspended growth process
- □ Biological nutrient removal
- □ Aerobic attached growth process (including trickling filters and biomembranes)
- □ Anaerobic attached growth process
- □ Pure oxygen aeration
- □ Aerated lagoon
- □ Facultative lagoon
- □ Other __________________________

18. Biosolids Processing Capacity at the study facility: _______ dry tons per day

19. If the study facility has both primary and secondary clarification, are the primary and secondary sludges processed separately?

- □ Yes, primary and secondary sludges are completely or partially processed separately
- □ No, all sludges are processed together throughout the biosolids treatment systems
- □ Not applicable

20. What sludge/biosolids treatment processes does the study facility perform? (check all that apply)

Preliminary	**Thickening**	**Stabilization**	**Dewatering**	**Conditioning**	**Heat Drying**
□ Grinding □ Screening □ Degritting □ Blending □ Storage □ Other ________	□ Gravity thickening □ Flotation thickening □ Centrifugation □ Gravity-belt thickening □ Rotary drum thickening □ Other ________	□ Alkaline stabilization □ Anaerobic digestion □ mesophilic □ thermophilic □ autothermal □ Aerobic digestion □ mesophilic □ thermophilic □ autothermal □ Composting □ Other ________	□ Centrifugation □ Belt-filter press □ Filter press □ Sludge drying beds □ Reed beds □ Lagoons □ Other ________	□ Chemical □ Other methods	□ Direct dryers □ Indirect dryers

21. Which of the following best describes the employee responsible for collecting biosolids samples from the study facility to test for microbes/pathogens?

- □ Plant superintendent
- □ Plant operator
- □ Professional technician
- □ Other

22. Does your utility require professional certification (e.g., operators certification) of the employees who collect biosolids samples for microbial/pathogen analysis?

- □ Yes
- □ No

Questions 23 through 25 concern the training and supervision that an employee receives specifically related to the collection of biosolids samples for microbial/pathogen analysis.

23. How much training (length of time) on sample collection does this employee receive before he or she takes samples without supervision?

    ____

24. Training for this task includes:
    (check all that apply)

    □ A video
    □ A manual
    □ A chart showing all the steps, posted where the samples are taken
    □ A chart showing all the steps, in a pocket guide
    □ Demonstration by a manager
    □ Demonstration by another non-management employee
    □ Supervised sampling by the employee, and the employee is supervised while performing this task ________ times. (Please write in number of times he or she performs task with the trainer present).
    □ Professional workshops
    □ Other (please describe)____

    ____

25. Is this employee subject to periodic review or observation about the collection and handling of biosolids samples for microbial/pathogen analysis?
    □ Yes
    □ No

26. Do you use any guidance documents to guide sampling methods? Please identify specific documents, including any guidance from state or federal regulators, professional publications, or Internet sources.

    □ Yes, we use the following guidance documents:

    □ No, we do not use any guidance documents.
    □

27. Are there any issues that you have concerning biosolids sampling and handling that you would like us to examine or would desire guidance for?

28. In a subsequent stage of the study we will be visiting a number of utilities in order to take samples and evaluate different sampling methods for a variety of biosolids products. If you'd like your treatment facilities to be considered, please provide contact information below.

- □ Yes, we would like for our facilities to be considered for the next stage of your study. You may contact us here: (Please provide the name and phone number of the person to contact or enclose a business card):

- □ No, we do not wish to participate.

29. May we follow-up with you to obtain a copy or reference to any guidance materials you use in order to help other information gathering aspects of this WERF Project?

- □ Yes, you may contact us here: (Please provide the name and phone number of the person to contact if you haven't done so already or enclose a business card):

- □ No, please do not contact us.

Thank you for participating in our survey. Please return the completed manager survey to us in the preaddressed stamped envelope within 15 days of receipt.

→ *Please write the name of the study facility on the enclosed envelope marked "sampler survey." (We ask this to ensure that your sampling staff member knows which facility we are studying. The envelope may be discarded after the survey is completed, so we will not see it.) Please give this envelope to an employee who regularly collects biosolids samples from the study facility.*

A-2 Sampler's Survey

UMBS-###s

Survey of Biosolids and Sludge Sampling Practices

To be completed by an employee who regularly collects biosolids and sludge samples for microbial analysis.

All of the questions in this survey refer to biosolids and sludge sampling at the facility whose name your manager has written on the envelope that contained this questionnaire. If your utility has more than one facility, please answer the questions as they relate to the facility identified on the envelope only.

1. Which of the following best describes your job classification?
   - □ Plant superintendent
   - □ Plant operator
   - □ Professional technician
   - □ Other

How many other employees besides yourself collect biosolids and sludge samples at this facility?

____

3. What types of biosolids materials are produced as end products at this facility? (check all that apply)
   - □ Cake (including granular)
   - □ Liquid
   - □ Compost
   - □ Other (please describe) ____

Questions 4 through 11 concern the training and supervision that you received specifically related to the collection and handling of biosolids and sludge samples for microbial/pathogen analysis.

Does your state require trained and certified individuals to sample and measure field test parameters like temperature, pH, and DO?

- ❑ Yes
- ❑ If yes, are you certified?
- ❑ No

How much training (length of time) on sample collection did you receive before taking samples without supervision?

Training for biosolids and sludge sampling included: (check all that apply)

- □ A video
- □ A manual
- □ A chart showing all the steps, posted where the samples are taken
- □ A chart showing all the steps, in a pocket guide
- □ Demonstration by a manager
- □ Demonstration by another non-management employee
- □ Supervised sampling by you, and you were supervised while performing this task ________ times. (Please write in number of times you performed the task with the trainer present).
- □ Professional workshops
- □ Other (please describe)____

Are you subject to periodic performance review or observation concerning the collection and handling of biosolids and sludge samples?

- □ Yes. If so, approximately how frequently are you reviewed or observed concerning the collection and handling of biosolids and sludge samples *(e.g.*, weekly, monthly, annually, etc.)?_________________
- □ No.

How adequately do you feel you were trained to collect and handle biosolids and sludge samples?

- □ Excellent training – I understood the importance of proper sample collection and handling, and I knew exactly what to do.
- □ Good training
- □ Fair training
- □ Poor training
- □ No training at all

Please list any aspects of sample collection and handling you would like to have been better informed about:

Do you have any other suggestions about ways that training could be improved?

Do you use any guidance documents to guide sampling methods? If so, please identify specific documents, including any guidance from state or federal regulators, professional publications, or Internet sources.

□ Yes, we use the following guidance documents:

□ No, we do not use any guidance documents.

Please list the major processes used to treat your sludge/biosolids to produce the final product and indicate whether you sample for microbial analysis before, during, and/or at the end of each process. For example, Process #1 might be "rotary drum thickening", Process #2 might be "thermophilic aerobic digestion", Process #3 might be "belt filter press", and so on.

Process #1______________________________
□ Biosolids/sludge samples are collected for microbial analysis □ before process □ during process □ at the end of process
□ Biosolids/sludge samples are not collected for microbial analysis in this process

Process #2______________________________
□ Biosolids/sludge samples are collected for microbial analysis □ before process □ during process □ at the end of process
□ Biosolids/sludge samples are not collected for microbial analysis in this process

Process #3______________________________
□ Biosolids/sludge samples are collected for microbial analysis □ before process □ during process □ at the end of process
□ Biosolids/sludge samples are not collected for microbial analysis in this process

Process #4______________________________
□ Biosolids/sludge samples are collected for microbial analysis □ before process □ during process □ at the end of process
□ Biosolids/sludge samples are not collected for microbial analysis in this process

Process #5______________________________
□ Biosolids/sludge samples are collected for microbial analysis □ before process □ during process □ at the end of process
□ Biosolids/sludge samples are not collected for microbial analysis in this process

Process #6______________________________
□ Biosolids/sludge samples are collected for microbial analysis □ before process □ during process □ at the end of process
□ Biosolids/sludge samples are not collected for microbial analysis in this process

Do the final biosolids products from your plant undergo additional treatment after they leave your facility?

- ❑ Yes
- ❑ No
- ❑ Not sure

The following questions concern testing for microbes/pathogens before biosolids products leave your facility.

Approximately what is the volume of material that you sample from when collecting sludge/biosolids for microbial/pathogen analysis? (*i.e.*, what is the size of the batch of material that you sample from?)

Cake volume (*e.g.*, how big is the pile?):____________________

Liquid volume (*e.g.*, how big is the tank?):____________________

Compost volume: (*e.g.*, how big is the pile?)____________________

What equipment do you use to take sludge/biosolids samples for microbial/pathogen analysis? Please describe. Please indicate whether the equipment you use requires any preparatory steps before sampling.

Cake sampling equipment:____________________

Liquid sampling equipment:____________________

Compost sampling equipment:____________________

How many samples do you collect from each batch of biosolids and how frequently do you collect samples?

Material	Number of sample drawn from a batch	Frequency of sampling
Cake		
Liquid		
Compost		
Other___________		

Do you store each sample in a separate container, or composite them in a common container?

- □ Each sample from the batch is stored in a separate container
- □ All the samples from the batch are placed in a single container
- □ Other (please explain)________________________________

Please describe in as much detail as possible how you take the samples (please complete for each sludge/biosolids matrix sampled as indicated below).

Compost (e.g., where do you take samples from?, are the contents of the compost pile mixed during sampling?, are samples taken a certain specific depth in the pile?, are there any other preparatory steps taken before collecting the actual sample?, etc.):

Liquid (e.g., where do you take samples from?, are the contents of the tank, etc. mixed during sampling?, are samples taken a certain specific liquid depth from the tank?, are there any other preparatory steps taken before collecting the actual sample?, etc.):

Cake (e.g., where do you take samples from?, is the cake broken up or mixed somehow before sampling?, are there any other preparatory steps taken before collecting the actual sample?, etc.):

What types of containers do you store the samples in?

Material	Container Material	Container Size	Opaque Container? (Yes/No)	Presterilized Container? (Yes/No)
Cake				
Liquid				
Compost				
Other________				

Comments
(optional):____

Please describe how you fill the containers. (What preservative do you use, how much headspace is left, do you do anything to the samples before sealing them, etc.)?

Material	Which preservative is used, if any?	How much headspace is retained?	Other handling/filling comments
Cake			
Liquid			
Compost			
Other________			

Comments
(optional):____

Where are sludge/biosolids samples usually analyzed for microbes/pathogens?

- □ At the utility
- □ They are shipped to an outside lab.
- □ Other______________

How long are sludge/biosolids samples typically held before shipping or onsite microbe/pathogen analysis?

Where are sludge/biosolids samples typically stored before they are shipped or analyzed onsite for microbes/pathogens?

- □ In a refrigerator
- □ In a cooler on ice
- □ Indoors at room temperature
- □ Outdoors at air temperature
- □ Other____

24. Are there any issues that you have concerning biosolids sampling and handling that you would like us to examine or that you would desire guidance for?

Thank you for participating in our study. Please return your completed survey form in the preaddressed envelope.

APPENDIX B

COMMUNICATIONS

B-1	*News & Notes*
B-2	*Communications Update to Facility Staff*
B-3	*Bulletin to Community Leaders*
B-4	*Technical Summary*
B-5	*Sample Guidance Poster*
B-6	*Compliance and Health Communication*
B-7	*One Page Sampling Instructions*
B-8	*Update to Research Community*

NEWS & NOTES

Communicating Issues about Biosolids Sampling

Operators-Managers Quarterly Technical Bulletin June 2007

As operators and managers responsible for overseeing all aspects of our wastewater treatment facility, there are numerous issues that require diligent attention—including the protocols and practices we employ for sampling biosolids products—either in the form of liquid, cake, or compost material.

What are primary issues concerning biosolids sampling activities within this wastewater utility we wish to effectively communicate?

KEY MANAGEMENT CONSIDERATIONS

1. To *effectively* maintain regulatory compliance and implement end-product quality controls it is necessary to also account for variability in treatment processes, staffing experience, and budget constraints.
2. Making sure that appropriate levels of training and oversight for quality control are in place (including potential certifications that may be required for certain employees) requires communication of the benefits to the overall operation of having these assurances in place.
3. Regularly assess budget needs and opportunities for providing training, outsourcing services (including laboratory expertise), acquiring proper equipment, and implementing compliance oversight.
4. Understand the implications of shifts in flows (feedstock and product demand), end-user needs, changing levels of process treatment requirements and changes in processing techniques.
5. Ensure consistent guidance resources are available from regulatory, research, and private services sectors and provided to appropriate operations personnel.
6. Maintaining public confidence in our operations.

As managers and operators, it is beneficial to communicate these specific management concerns to other key stakeholders.

IDENTIFYING COMMON QUALITY CONTROL ISSUES IN MANAGEMENT

ITEM 1

Effectively maintaining regulatory compliance and quality controls given variability in treatment processes, staffing experience, and budget constraints

- Highly diverse wastewater treatment processes are in place and can influence biosolids quality characteristics at various points within the treatment process. With this in mind it is possible certain sampling practices that are necessary for one process would not necessarily be appropriate for another (e.g. anaerobic vs. aerobic). Consistency in treatment process biosolids outputs, if compromised, requires additional assurances in sampling practices are in place.
- Staff turnover and shifts in employee responsibilities can result in situations where individuals taking samples lack sufficient training, experience or understanding about the importance of proper protocol.
- Product quality management can be influenced by end-user markets and vendor relationships. End-users that are informed and are utilizing products in higher valued applications may expect more evidence that quality controls are in place.
- Budget limitations often influence prioritization of expenditures dedicated to quality controls recognizing investments required for treatment processes and broader sampling needs.

ITEM 2

Ensuring health and safety training and oversight for quality control, including certifications required by staff, and communicating the associated benefits.

- Clearly it is advantageous to the operation to retain well trained and knowledgeable staff for all aspects of operations. The level of health and safety training individual employees obtain may vary based on their position as well as government agency specifications.
- Different oversight agencies or levels of government (local, state, federal) may require certifications yet they may not always be consistent with each other. This must be recognized by management.
- Training effectiveness may be impacted by the source of training (internal versus external) and the type of training (written documents, workshop, video and/or on-site demonstrations). Oversight and proper documentation are expected to influence the application of training and protocols.

ITEM 3

Assess budget constraints and associated opportunities for providing training, outsourcing services (including laboratory expertise), obtaining proper equipment, and compliance oversight.

- There are cost-benefit trade-offs to consider with all investments that impact quality assurance, health and safety compliance, and operational efficiency.
- Available budgets will influence investments in training and certification of employees as well as the level of on-going training samplers and technicians receive.
- Availability of resources influences the ability to hire highly skilled employees or to outsource assistance with sampling and analysis. Funding can influence the decision to employ on-site or off-site laboratory analysis.
- The costs of obtaining and maintaining appropriate equipment for sampling practices is a consideration in the development of a viable sampling program.
- Implementation of consistent, functional compliance practices requires commitment to oversight and procedures for handling of situations where responsible practice is compromised.
- Aligning economic considerations with product quality demands requires careful budgeting for:
  - sample collection and analysis;
  - processing and treatment upgrades (in relationship to consistent product quality production); and
  - meeting all forms of operational compliance besides sampling for microbial analysis quality control

IDENTIFYING COMMON QUALITY CONTROL ISSUES IN MANAGEMENT

(CONTINUED)

ITEM 4

Implications of shifts in flows (feedstock and product demand), end-user needs, levels of treatment required and processing techniques

- The quality of biosolids may be influenced by shifts in influent or solids (feedstock) flows and associated characteristics. Such shifts could have implications for continuing previously accepted sampling protocols.
- Changes in product demand and end-user requirements will influence the throughput of biosolids products and thus can influence sampling demands.
- Changes in the level of wastewater treatment and sludge processing, if changed, may result in variable output volumes and end-product characteristics.

ITEM 5

Ensure consistent guidance resources are available from regulatory, research, and private services sectors and provided to appropriate operations personnel.

- Management can best coordinate employee training needs, cost effective compliance practices, and oversight other operational needs including sampling activities when credible, consistent, and clear guidance resources are available.
- States having specific certification requirements and clear guidelines for microbial analysis can influence operation decisions and the choice to implement QA/QC programs within facilities.
- Laboratories and professional service providers can be good sources of assistance however consistency in recommendations, guidelines, and practice must be ensured.
- Templates useful for creating effective QA/QC documents, SOPs, and internal guidelines for sampling could be a benefit to managers.
- Research informing the development of practical, science-based sampling protocols that can be both consistently applied within the industry and utilized by different types of wastewater facilities should be available to management.

Management can create specific text box content here such as referencing their QA/QC document, note about their processes, sources of guidance they use, etc. Management can create specific text box content here such as referencing their QA/QC document, note about their processes, sources of guidance they use, etc. Management can create specific text box content here such as referencing their QA/QC document, note about their processes, sources of guidance they use, etc. Management can create specific text box content here such as referencing their QA/QC document, note about their processes, sources of guidance they use, etc.

COMMUNICATION NEEDS AS MANAGERS

Internal Communications Needs

- With staff and employees to help them to understand the importance of safety and quality parameters and related environmental and human health concerns
- Recognized importance of each employees contributions to the operational success of the facility and the pride they should take in their contributions to quality assurance
- Sharing how the facilities bottom-line can be affected by maintaining good practices and quality biosolids materials for beneficial use
- Practitioners, technicians and administrative staff all need to understand what protocols are in place within the operation, which are required, and why they are important

External Communications Needs

- Informing regulatory and public health professionals of concerns with protocols, guidance, compliance challenges and other tools useful to the operation
- Increase awareness of QA/QC steps to end-users, consumers, and vendors that demand various types of biosolids products
- Notifying community leaders of the public benefits resulting from the biosolids processing and having proper sampling protocols in place
- Apprising engineers, researchers, and professional service providers of the impacts different treatment process designs and implementation have on facility performance (technical and operational)

BENEFITS OF COMMUNICATIONS

The overall technical and economic performance of the operation can be expected to improve as communication about different issues expands and is prioritized. This can foster integration of practical experience with policy and research all aimed at improving strategies for health and safety compliance. When stakeholders are given an opportunity to better understand the issues specific to diverse groups, public relations and professional integrity benefit. Thanks for reading NEWS & NOTES.

B-2 Communications Update to Facility Staff

[RecipientName]
[Address Max. 6 lines]

Facility Name
Facility Address
Facility City
Facility State and Zip
Tel xxx.xxx.xxxx
Fax xxx.xxx.xxxx
www.FACILITY.com

Subject:
Communications Update to Facility Staff

Date:
[MonthDay,Year]

Contact:
[Contact]

Phone:
[Phone No.]

Email:
[Email Address]

Our ref:
[Project No.]

Dear Facility Staff:
Quality Control practices are extremely important to the success of our wastewater operation and to the way our facility is viewed by the customers we serve. Managing our biosolids to the standards our customers and oversight agencies expect requires diligence and pride in our work as a team. As your facility manager it is my goal to communicate with everyone specific issues that impact us all. I hope you find this edition of Staff Update. **(managers encouraged to create their own unique title for this piece)**

Biosolids Sampling Protocols

Our Compliance Needs

Here in (utility fill in City, State) we are subject to the following agency compliance requirements specific to how we sample our biosolids materials for analysis of microbial content. These guidelines are updated every (utility fill in timeframe or indicate how often subject to review). If guidelines have similar requirements for sampling parameters other than microbial content I have indicated so below.

1. The State of (fill in, if no state requirements indicate this here) requires we utilize the following sampling guidelines for microbial analysis:
    a. Example
    b. etc.
2. We primarily rely on the following (federal, laboratory, etc) guidance resources
    a. (Reference guidelines here)
    b. insert guidance type
    c. etc
3. We (if appropriate, insert require or are required) staff responsible for sampling to obtain (insert type) certification.
4. Our operation also has written guidance on up-to-date sampling protocols in our facility (insert what is appropriate here such as QC/QA program, EMS plan, SOP, etc).

Regardless of the guidance protocol applied, the following practices are unequivocally important to apply when collecting samples for microbial analysis.

- Always use sterilized and autoclavable (if possible) containers to collect samples
- Always place sample containers in coolers using blue packs, gel ice packs, ice and/or ice water bath or refrigerate ***immediately*** after collection
- If sampling from pipes, always clear spigot of dried or cake material; waste the first purge of biosolids material before collecting the sample for analysis; avoid touching outlet with sample container

Application of Clean Water Act "503 Rule" (Title 40 CFR, Part 503)

The rule contains requirements for the final use or disposal of biosolids. Your state may have more stringent regulations for biosolids in place than the Federal 503 guidelines. The 503 rule:

e) Requires sampling and analysis for pollutants (metals), pathogens, and vector attraction
f) Specifies appropriate analytical methods for these criteria
g) Advises on sampling frequency and monitoring
h) DOES NOT provide guidance on how to collect samples

The rule specifies the number of samples to be collected ONLY for fecal coliform monitoring of Class B, Alternative 1 biosolids.

The rule provides general guidance on sampling however more specific protocols required can be found in association with laboratory analytical methods guidance documents, more recently EPA Methods 1680, 1681, and 1682, and state regulations associated with the permits obtained by your facility.

Health and Safety

Not only is the protection of environmental and public health important to our operation so is the assurance that *you*, one of *our valued employees*, are engaged in practices that protect your health and well-being.

Recommendations on What to Wear

- Remember when collecting samples regardless of biosolids type:
    - Use protective gloves and dispose of properly
    - Wear protective eye gear
    - Use a mask, particularly when aerosols may be present
    - Wear shoes that will not be worn into non-storage or public areas

- Change clothes in the event they become soiled with biosolids sampled

- When working around machinery and processing equipment wear:
    - Gloves where necessary, fitting tightly to the hand
    - Protective eye gear such as impact resistant glasses or goggles
    - Protective boots

Clean-up Options

- Prior to handling sampling equipment be sure to properly sanitize equipment following sampling protocols

- After collecting samples the following can be disposed of as follows
    - (**insert specifics here** .. for rubber gloves, goggles, etc)

- These materials should be washed using
    - (**insert here type of soap for clothes, rags, hats, etc**)

- Upon completion of all sampling activities use at least (**type of soap, warm or hot water**) to thoroughly wash hands, neck, and face. Showers are available for all employees to use as deemed necessary if an exposure incident of concern is encountered.

- (**Perhaps insert something about handling rooms, etc and appropriate wash down if this in fact exists**)

Signals and Signs

If you:

- become nauseous

- experience severe watering, burning, or itching of eyes

- have severe itching or burning of the skin or a noticeable breakout of a rash

- experience difficulty breathing or maintaining a clear breathing pattern

Contact either myself or your immediate supervisor immediately. You may also contact a health services line by calling (insert a number here), text messaging (insert if available here), or e-mailing (proper e-mail address here). Be sure to explain the best you are able the most recent work activity you participated in and details of the symptoms experienced.

Training Updates

The following training requirements are in place for all employees involved in sampling biosolids materials.

(Utility specifically insert what is relevant here specific to the facility, state or some oversight body)

New training resources available include (for whichever ones a facility can list would suggest a short one or two sentence description of each be included):

- video
- new pocket guide
- web based source
- workshop

Know the Customer

Our customers appreciate knowing we provide quality biosolids material for their use. We currently provide material to these end-users

(Utility specifically insert who customer is – ag, landscape, forestry, restoration)

Customers in the (xyz sector) require our materials meet the following microbial analysis criteria and thus the reason we must be diligent in following sampling protocols.

Pride in What You Do

Taking pride in the quality of your work as samplers for this operation is important to our utility and most importantly to the community we serve. Without the level of attention to good work you provide we would not have the public support and consumer demand for our biosolids materials. Even the smallest detail in sample collection and handling can make a difference in protection of environmental and public health. Consider your value in this cause each and every time you fulfill this task ... **WHAT YOU DO MATTERS** ... don't hesitate to communicate with management your needs and concerns.

Sampler TIP of the MONTH

Utility specifically insert a sampling tip or have a sampler contribute a lesson learned here.

Sincerely,

[Author Name]
[Author Title]

Copies:
[Copies]

B-3 Bulletin to Community Leaders

Community Update

Bulletin

Communicating Biosolids Sampling Issues to Regulators & Community Leaders

Practice and Protocol

Sampler Training **{utility insert what is practiced, formal and/or informal types}**

To ensure health and safety protocols are met we sample biosolids materials for all regulated criteria and give particular attention to microbial analysis requirements. There are several ways we can ensure quality controls are practiced.

- Practitioners are required to **{utility insert the form of training or certification necessary here}**
- Samplers are supervised by management and experienced staff during a minimum of **{utility insert number}** of sampling events per year until demonstrated expertise is recognized. **{utility should word this in accordance with their practice if differs the current statement}**
- Samplers are provided with on-site, readily accessible references to proper sampling protocols.
- Samplers have an open door policy for communicating health and safety concerns to management.
- **{If sampling activity is contracted to outside professionals utility insert here if so and list who}**

Equipment and Tools

- All sampling technicians and practitioners are provided with designated and appropriate equipment for the specific biosolids material to be sampled
- Sampling tools are properly sterilized **{utility insert example of what is practiced}** and safely disposed of when no longer appropriate for use
- When possible equipment provided to samplers is standardized for collection and handling of appropriate sample volumes
- Samples are stored in approved containers and handled in accordance with approved regulatory protocols and/or laboratory specifications

Communication

Our wastewater operation implements a communications strategy that helps to keep all employees regardless of position informed about our compliance and performance requirements as well as our responsibility to the community. **{If applicable insert – As part of our EMS program different communication efforts are implemented}** We provide updates to regulatory and public health professionals and encourage open exchange of ideas with these and other relevant stakeholders concerning how to improve the efficiency and effectiveness of our operation.

Guidelines and Protocols

- Our facility follows (**utility insert regulatory or agency specific guidance references here**) for biosolids sampling and handling
- This operation has a **(QA/QC plan, EMS program, SOP, fill in as appropriate)** in place specific to our facility that does include of guidance specific to biosolids sampling activities
- Management attends seminars and (**utility insert specifics here such as webcasts, workshops, etc and frequency**) to remain up-to-date on accepted sampling protocols
- (**If applicable utility insert certification steps or continuing education credits staff undertakes important to sampling responsibilities**)
- Our facility communicates directly with the laboratory doing analysis to obtain any additional specific sample collection and handling requirements

Continuing quality public service to our community also requires diligent practices be maintained and open communication occurs with our community leaders.

B-4 Biosolids Sampling for Microbial Analysis – Appropriate Practices for Quality Assurance and Quality Control (QA/QC)

Biosolids Samplers and Practitioners – Do you often wonder WHAT protocols you are *really* supposed to follow and occasionally wonder WHY?

WHY SWEAT THE DETAILS?

Reasons for QA/QC Practices

What are the *reasons* quality assurance and quality control practices specific to biosolids sampling are required by your wastewater treatment facility?

- Your facility is subject to federal, state, and/or local regulations and guidelines that require the biosolids materials produced meet certain standards for potential contaminants such as metals, nutrient, and pathogen and/or indicator content.
- The protection of human and environmental health from exposure to potential contaminants in the biosolids material demands that various safety measures be in place.
- Risks and liabilities to your facility can be mitigated by applying best practices and quality controls, particularly in the event unexpected exposures occur.
- The biosolids produced at your facility may be considered a valuable resource intended to be put to beneficial use as a fertilizer, soil amendment, soil stabilization material, etc. Biosolids products could be utilized for:
  - Agriculture
  - Landscaping and horticulture
  - Forestry
  - Impaired property closure (landfills, mines)

Practicing consistent QA/QC procedures protects public health and the environment, and results in better public relations, consumer confidence, and overall performance of your wastewater facility.

WHAT SHOULD BE PRACTICED?

Knowing and Understanding Accepted Sampling Protocols

Applying *accepted practices and protocols* for collection and handling of biosolids samples is guided by:

- Regulatory compliance requirements
- End-product user requirements
- Type of laboratory analysis to be performed.

Accepted practices for biosolids sampling may be influenced by geographically specific guidelines and treatment processes and methods of storing biosolids. Sampling practices for metals and nutrient analysis typically differs those for pathogen indicators and microbial content. The protocols in place for microbial and pathogen indicator sampling may be very similar but will be site and end-use specific.

Regardless of where you are located and the processes used to produce your biosolids product(s), it is important that you, your management, and the laboratory responsible for sample analysis fully ***communicate about****:*

1) How the samples should be collected

- How frequently should sample(s) be collected and how many samples per sampling period must be obtained?
- Where in the material flow, treatment process, or stored volume should the sample(s) be collected?
- What is the size (volume) and number of sample(s) that should be collected?
- What equipment should be used to collect the sample(s) and how should this equipment be maintained?

2) How the samples are to be handled and transported to the laboratory facility for analysis

- What size and what type (material) container is to be used?
- Are any additional sterilization procedures required?
- What are the temperature requirements for the stored sample(s)?
- What are the maximum holding times for sample(s) and does it vary with storage temperature?
- In collecting samples, are individual grab (discrete) or composite samples to be submitted to the laboratory?

3) Do sample collection practices need to be specified for biosolids material (cake, liquid, compost) type only or does biosolids treatment process also need to be considered?

4) What are the available guidance resources describing accepted sampling protocols and which available source should be prioritized?

DEVIL IN THE DETAILS - TECHINCAL POINTS [1]

Sample Collection

1. Sampling *frequency* is dependent on the total metric tons of biosolids that are produced by your facility. Select your operations capacity and required frequency as follows:

Metric Tons Produced		Collection Frequency
< 290 mt	↔	1X/yr
290 to 1500 mt	↔	4X/yr
1500 to 15,000 mt	↔	6X/yr
<15,000 mt	↔	12X/yr

Note: Collection of samples should be as close as possible to the end of the treatment process or to the time the biosolids are to be land applied.

2. The specific sampling *location* depends on the type of biosolids material produced and should be taken at a point where any final mixing, if applicable, occurs:
   - For **liquid** – samples can be obtained from a side spigot or a transfer pipe (purge pipes to clear dried or caked materials prior to collection); option is to directly collect from tanks
   - For **cake** – collect from an end-of-process conveyor or a storage pile
   - For **compost** – collect from piles at the completion of the "active composting" phase for agricultural uses and after the prescribed curing period for landscape uses

3. Sample collection *volume* and sample *replication*:

   MORE IS NOT BETTER!

   - EPA Methods[2] specify filling **1 liter** sample containers
     - Liquid samples require leaving 1/2" headspace
     - Compost or cake do not require headspace and can be put into larger containers however 1-liter volume is still advised
   - Collection of multiple discrete samples is recommended for determining product variability. Use of discrete (grab) samples is typically specified for analysis of pathogen or microbial indicator content. Some NPDES permits may specify either composite or grab samples are accepted.
   - For Class B, Alternative 1 biosolids, <u>7 discrete (grab) samples</u> must be analyzed. A greater number of samples should be collected and analyzed when the facility is producing larger quantities of biosolids.

---

[1] The following guidelines are synthesized from existing guidance (503 Regulations and associated EPA Reports) and the results from WERF Project 04-HHE-7.

[2] EPA Methods in reference include 1680, 1681, and 1682.

> **Application of Clean Water Act "503 Rule"**
> **(Title 40 CFR, Part 503)**
>
> The rule contains requirements for the final use or disposal of biosolids. Your state may have more stringent regulations for biosolids in place than the Federal 503 guidelines. The 503 rule:
>
> a) Requires sampling and analysis for pollutants (metals), pathogens, and vector attraction
> b) Specifies appropriate analytical methods for these criteria
> c) Advises on sampling frequency and monitoring
> d) DOES NOT provide guidance on how to collect samples
>
> The rule specifies the number of samples to be collected ONLY for fecal coliform monitoring of Class B, Alternative 1 biosolids.
>
> The rule provides general guidance on sampling however more specific protocols required can be found in association with laboratory analytical methods guidance documents, more recently EPA Methods 1680, 1681, and 1682, and state regulations associated with the permits obtained by your facility.

- For facilities that store biosolids for an extended period of time such as greater than 30 days additional sampling and testing of stored material should be conducted prior to land application. The **total number** of samples required is dependent on facility wastewater design flows and can be determined using the following reference:

Facility Design Flow	**Sampling Replication**
< 1 MGD	27 discrete *into* 1 composite
1 to 10 MGD	42 discrete *into* 4 composites
10 to 100 MGD	48 discrete *into* 6 composites
> 100 MGD	84 discrete *into* 12 composites

4. Sample *bottles* should be comprised of autoclavable plastic (polyethylene, polypropylene, teflon), glass or stainless steel.

5. Proper sampling *equipment* is dependent on the physical characteristics of the biosolids and how the biosolids material is being stored. Equipment made of plastic, steel and stainless steel is generally acceptable. *Equipment that is galvanized and may contain zinc should never be used.* Other considerations for selecting proper sampling equipment include temperature of compost at sampling location in the pile; volume or mass and configuration of biosolids that needs to be sampled. Standard guidance is as follows:

Material Type	Device	Size	Type Sample
Liquid			
- free-flowing	Liquid composite	5-15' long	Composite
- other?	Dipper or bucket	1 liter, 5-15' long	Composite
	Bottle	1 liter	Grab/discrete
Semi-solid	Trier/Thief sampler Shovel or scoop	4 ft or less	Discrete or cross-section
Loose-solids	Shovel or scoop	variable	Grab/discrete
Hard-packed solids	Auger	variable	Discrete or cross-section

6. Equipment *preparation*:
   First time use
   - Wash, wrap and autoclave when possible

   Secondary use
   - When sampling devices such as dippers or augers or shovels need to be used to collect a number of discrete (grab) samples field sterilization techniques can be applied including
     – washing all debris or cake material from devices with soap and tap water
     – rinsing all surfaces three times with 50% ethanol
     – rinsing all surfaces three times with autoclaved laboratory water
     – properly reuse sampling device

Handling and Transport

All appropriate and accepted guidance documents should reference cooling and storage times in relationship to the microbial analysis to be performed. Of the most frequently cited guidance documents cited[3], the following are consistently required:

Analysis	**Holding Time**	**Temperature**
Fecal coliform *Salmonella*	24 hrs[4]	4°C
Enteric viruses	≤ 2 weeks	frozen; 0-18°C
Helminth	1 month	4°C

For all analyses holding times recommended of 1-6 hours also required < 10°C temperature and 4°C during shipping.[5]

[3] The White House Document, Plain English Guide, and EPA 831B-93-002a

[4] For Class A compost, Class A or B digested biosolids, for all others within 6 hours is advised

[5] Inclusive of guidance documents EPA833-B-89-100 and MOP-OM-1

For Onsite Analysis

- An ice water bath provides the most rapid cooling and is preferred. *Exercise care to avoid cross- contamination of samples with ice water.*

For Offsite Shipping

- Pack bottles well surrounded by blue ice packs or gel ice packs that has been stored in a -20°C freezer for at least 24 hours
- An additional temperature control sample should also be shipped making sure the center core temperature of this sample is below 10°C upon destination arrival.
- Hand delivery of samples within 6 hours is preferred or use express overnight shipping with morning delivery having samples collected as late in the day as possible to meet shipping deadlines
- Samples can be placed in compact sized refrigerators only if they can be analyzed within 6 hours

Useful References

U.S. EPA. 1989. POTW Sludge Sampling and Analysis Guidance Document. EPA 833-B-89-100, Office of Water, Washington, D.C. Available through the internet: http://www.epa.gov/npdes/pubs/owm012.pdf.

U.S. EPA. 1993. Preparing Sewage Sludge for Land Application or Surface Disposal. EPA831B/93/002a, Office of Water, Washington, D.C. Available through the National Service Center for Environmental Publications (NSCEP): http://www.epa.gov/nscep/.

U.S. EPA. 1994. Plain English Guide to the EPA Part 503 Biosolids Rule. EPA 832/R-93/003, Office of Wastewater Management, Washington, D.C.

U.S. EPA. 2003. White House Document (Revised July 2003) EPA 625/R-92/013, available through the EPA website: http://www.epa.gov/ord/NRMRL/Pubs/1992/625R92013.pdf.

U.S. EPA. 2006. Method 1680: Fecal Coliforms in Sewage Sludge (Biosolids) by Multiple-Tube Fermentation using Lauryl Tryptose Broth (LTB) and EC Medium, EPA-821-R-04-026; EPA Office of Water: Washington, D.C.

U.S. EPA. 2006a. Method 1681: Fecal Coliforms in Sewage Sludge (Biosolids) by Multiple-Tube Fermentation using A-1 Medium, EPA-821-R-06-013, EPA Office Water: Washington, D.C.

U.S. EPA. 2006b. Method 1682: *Salmonella* in Sewage Sludge (Biosolids) by Modified Semisolid Rappaport-Vassiliadis (MSRV) Medium, EPA-821-R-06-014, EPA Office of Water: Washington, D.C.

B-5 Sample Guidance Poster

Biosolids Sampling

This poster is formatted to print 24" x 36". Larger sizes may be possible.

Microbial Sampling Locations

Influent Process

⇩

Digestion

⇩

Effluent Stream

⇩

Sampling Point

ALWAYS attempt to collect your sample as near to the end of the treatment process as possible.

This means that for…

Liquid

- Collect samples after purging pipes to clear any hardened material and collect from spigot or pipe outlet
- Collect after storage tank has been thoroughly mixed at a location 50-75% of tank depth

Compost Pile

- For biosolids used in agriculture sample piles when active phase is complete; for landscape utilization sample after prescribed curing phase
- Collect samples at least 6 inches deep into pile and use an appropriate grid pattern relative to your pile configuration

Cake

- Sample as material on conveyors leaving dewatering device going to storage
- Collect when biosolids have stabilized

WERF Project 04-HHE-7

Principal Investigator
Sharon Long, PhD, University of Wisconsin*

Communications Consultant
Anita Bahe, PhD, Lynx Group International

*This project originated in conjunction with the University of Massachusetts

Sampling Tools and Equipment

BE SURE TO sterilize tools prior to use by autoclaving. IN THE FIELD, use soap, deionized water, 50-percent ethanol rinse, and autoclaved laboratory water.

Sample Scoops

- An all metal spoon that can be wrapped and autoclaved; only use a large curved surface with screened material
- Do not use a curved blade trowel with plastic handle
- Flat blade trowels with 6-inch blade can be wrapped and autoclaved; can use to ensure at least 6 inches of compost surface is removed

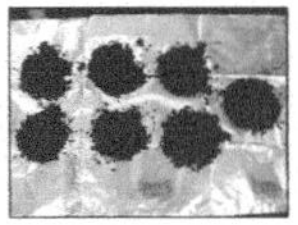

Preparing Composites

If composite samples are acceptable, subsample using 150-mL or 4 oz. specimen cups and mix on autoclaved aluminum foil.

Cake/Compost Sampling

- Collect 1-liter samples; a 1-liter bottle can be filled to capacity
- Prefer grab samples of cake; composite samples of compost

Liquid Sampling

- Proper sample containers consistent with EPA-approved microbial analysis methods are must be used. Items:

(1) Sterilized, **1-liter** wide-mouth sample bottle (autoclave indicator tape visible)

(2,3) Properly filled 1-liter bottles with at least 0.5-inch headspace

(4,5) Inappropriately large sample containers for 1-liter samples.

Liquid Sampling Kit

- Supplies required for collecting liquid samples for nutrients, metals, and microbials.

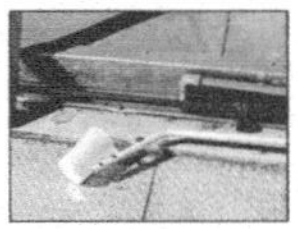

Dipper

Typical dipper attached to a 15-foot pole for collecting liquid samples from tanks. Field sterilization required.

Hand Pump

Hand pump sampling device that can be retrofitted for liquid sampling from tanks.

NOTE: Figures correlate from top to bottom.

Handling Samples

Vessel Size and Composition

- Prefer using ***1-liter*** containers; larger containers can be used for compost and cake but ***sample size must be 1-liter volume***
- Autoclavable plastic (polyethylene, polypropylene, Teflon), glass or stainless steel

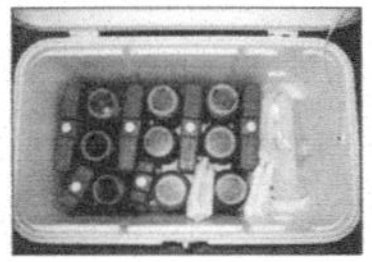

Packing Requirements

- Use blue ice packs, gel ice packs, ice or ice water
- Evenly distribute cooling materials around sample vessels
- See example photo (right)

Storage Temperature and Holding Time

- Fecal coliform, *Salmonella*: ***24 hours @ 4 °C***
- Enteric viruses: ***≤2 weeks @ -18 to 0 C***
- Helminth: ***1 month @ 4 C***

Frequency, Volume, and Number

Frequency of collecting samples is determined by the **metric tons** of biosolids produced annually at this facility.

Frequency	Metric tons
1 x per year	Up to 290 metric tons
4 x per year	290 – 1,500 metric tons
6 x per year	1,500 – 15,000 metric tons
12 x per year	15,000 metric tons or greater

- For all three types of biosolids (liquid, cake, compost) collect **1-Liter** quantities for analysis. When collecting **liquid** leave **½" headspace** in the sample analysis container.
- Collecting **discrete (grab)** samples is recommended for microbial content analysis. When producing "**Class B, Alternative1**" material, **seven (7)** discrete (grab) samples must be analyzed.
- For **biosolids stored longer than 30** days the number of samples to collect depends on the **design flow** of your facility.

Facility Design Flow	Sampling Replication
< 1 MGD	27 discrete *into* 1 composite
1 to 10 MGD	42 discrete *into* 4 composites
10 to 100 MGD	48 discrete *into* 6 composites
> 100 MGD	84 discrete *into* 12 composites

Laboratory for Analysis

Guidance Resources

- *U.S. EPA Methods 1680, 1681, 1682 (2006)*
- *Title 40 CFR, Part 503 EPA/625/R-92/013 (July 2003)*
- *U.S. Composting Council TMECC (Test Methods for the Examination of Composting and Compost)*

Shipping

Be sure proper manifests are in place and all shipping requirements are met. See supplemental sheet provided.

B-6 Compliance and Public Health Bulletin Update to the Research Community

Bulletin

Public Health and Compliance

Communicating Biosolids Management Practices for Protecting Public Health

In This Issue

Biosolids Quality Controls: Proper Sampling and Handling for Microbial Analysis

Your local wastewater utility provides a valuable service to the community. Nationwide a variety of pollutant and pathogen removal processes are used to treat wastewater from homes and businesses, each process resulting in the creation of different types of by-products including ***biosolids***.

With proper management and analysis these biosolids can be sustainably utilized as a sources of organic matter and nutrients for growing plants, as a material for capping landfills or reclaiming mines, and in some instances as a fuel source. Wastewater operations are subject to strict guidelines for pollutant and pathogen removal as governed by state and federal regulations and permits. Likewise end products of treatment processes (biosolids) are also subject to environmental and human health protection standards which are set, in part based on the planned end-use of these materials.

Biosolids quality compliance is assessed through testing. Assuring the quality and safety of biosolids requires proper sampling and handling protocols are in place and are practiced for characterizing the materials and in particular for analyzing microorganism and pathogen content. Testing is conducted to prove that treatment processes are reducing the potential for unhealthy conditions and unwanted environmental impacts to occur.

Types of Biosolids

Biosolids for beneficial end-uses are often in the form of ***Liquid***, ***Cake***, or ***Compost***. Biosolids intended for beneficial use may be classified as ***Class A*** or ***Class B*** as determined by the level of treatment processing and the resulting quality characteristics represented as pathogen indicator content.

Primary End-Uses

The most common uses are for:

- ***Agriculture***
- ***Landscaping and nurseries***
- ***Forestry***
- ***Impaired property closures (landfills, mines, etc.)***

Biosolids that are not put to beneficial use are typically disposed by landfilling or incineration.

What Are Biosolids?

Biosolids are the nutrient-rich organic materials resulting from the treatment of sewage sludge (the name for the solid, semisolid or liquid untreated residue generated during the treatment of domestic sewage in a treatment facility). When treated and processed, sewage sludge becomes biosolids which can be safely recycled and applied as fertilizer to sustainably improve and maintain productive soils and stimulate plant growth. Biosolids are typically sampled and analyzed for different constituents including nutrients, metals, and indicator organisms and/or pathogens .

Why is attention to sampling procedures important to understand?

The use of biosolids depends on collecting data that demonstrates acceptable quality has been met. Despite this need, concise and consistent guidance *specific to biosolids sample collection and handling practices for microbial analyses* are not always readily available to managers and samplers at utilities. Regulators and public officials need to be sure the best protocols are available and can be implemented by your local treatment facility.

Public Health and Compliance Professionals Can Make a Difference

Helping to identifying practical, science-based sampling protocols that are cost-effective and ensure that better quality controls are in place can result in greater consistency in the management of the biosolids. It is equally important to understand that the diverse nature of wastewater treatment operations and the variability of end-user expectations ultimately demands a trained staff be engaged in diligent sampling activities.

What are some practical considerations?

1. Specific wastewater and biosolids treatment processes yield biosolids materials with unique characteristics thus influencing requirements for obtaining accurate samples for microbial analysis. Adherence to guidance provided by analytical laboratories used to conduct microbial analyses may limit flexibility and may not always be practical for all types and sizes of operations.
2. Not all states impose the same requirements for sampling and/or personnel training and certification which can result in inconsistency in compliance. Variations in management oversight due to budget constraints can also result in different levels of quality control and quality assurance practices.
3. The diverse nature of treatment facility size and operation results in variability in resources for sampling equipment, training, and laboratory analysis services. Because of this situation simplicity without compromise to science-based guidance and practice can result in greater consistency and effectiveness in sampling activity.

Science and Technical Support

It is recognized that science and technical discoveries can continue to improve the effectiveness of sampling for biosolids reuse by better understanding:

- Which sampling methods and associated frequency of collection as related to the facility's treatment processes, and the size of the operation provide statistically sound representation of onsite biosolids materials
- What are the easy but effective ways to collect, store and ship biosolids to prevent misrepresentative samples

The Good News

1. Wastewater operations are more often creating comprehensive Environmental Management Systems and in-house Quality Assurance/ Quality Control plans to complement sampling guidance available from state and federal sources.
2. The demand for biosolids acceptable for reuse creates competition for providing quality, safe materials and therefore further investments in the resources necessary to meet those demands
3. Treatment technologies have improved over time resulting in higher quality by-products at end-of-process. Laboratory analysis methods are improving, as are the methods for collecting appropriate samples for specific types of analysis.
4. Communications are improving across all stakeholder groups, within operations, and across the industry thus helping to advance problem-solving and increase confidence.

Producing quality biosolids for beneficial utilization does not stop with treatment processes. Public confidence in safe and healthy reutilization of biosolids materials requires that consistent and science-based sampling and analysis protocols are in place.

B-7 One Page Sampling Instructions

Sampling Cake for Microbial Analysis

*Attempt to select sampling location **as near** the end of the treatment process as possible – further paying attention to:*

Preparation

- Sterilize scoops, shovels, dippers or other collection tools using heat sterilization (autoclaving) when possible;
    - In the field between samples was with soap and deionized water followed by a 50% ethanol rinse and a rinse with autoclaved laboratory water
- ***Do not*** use a blade trowel or scoop with a ***plastic handle*** as this device can not be autoclaved
- Sterilize autoclavable plastic, glass or steel sample storage containers
- If the agency administering your NPDES permit accepts composite samples autoclave aluminum foil to be used for composite sample preparation

Sampling

- Sample the cake on conveyors as it leaves the dewatering device going to storage
- Collect samples when the biosolids have been stabilized
- Fill a **1-liter** sample bottle to capacity - containers larger than 1-liter can be used but it is advised that no more than a 1-liter sample be obtained
    - If composite samples are acceptable to your permitting agency, 150-mL or 4 oz. sterile specimen cups can be used to collect subsamples for use in producing a composite sample; mix replicates on autoclaved foil prior to filling final sample container

Packing and Storage

- Place bottles in cooler so ice packs and ice water can be evenly distributed around containers – blue packs or ice packs are preferred
- Apply the following temperature and holding time recommendations:

Fecal coliform and *Salmonella*	4 °C	24 hours
Enteric viruses	<0 °C	up to 2 weeks
Helminth	4 °C	1 month

Shipping

- Hand deliver or send by courier on ice for analysis within 6 hours of collection if possible; or
- Collect samples in the afternoon, ship on ice via express mail with overnight, morning delivery

Be SURE to read the shipping instructions brochure and documentation requirements!

Sampling Compost for Microbial Analysis

Plan to collect your samples ***as near*** *the end of compost processing and curing as possible – giving further attention to:*

Preparation

- Sterilize scoops, shovels, dippers or other collection tools using heat sterilization (autoclaving) when possible;
  - In the field between samples was with soap and deionized water followed by a 50% ethanol rinse and a rinse with autoclaved laboratory water
- ***Do not*** use a blade trowel or scoop with a ***plastic handle*** as this device can not be autoclaved
- Sterilize autoclavable plastic, glass or steel sample storage containers
- Autoclave aluminum foil to be used for composite sample preparation

Sampling

- If biosolids are used for agriculture collect sample when the *active phase* of composting is complete
- When biosolids are used for landscaping collect sample after the *prescribed curing phase*
- Analysis of composite compost samples is allowed by EPA. To produce a representative composite sample:
  - Use uniform sized sterile specimen cups or jars (150 mL or 4 oz. containers) for subsamples
  - Composite by mixing subsamples on autoclaved foil or in a sterilized bucket with a sterile plastic bag
- Obtain enough material from the composite mix to fill a **1-liter** sample bottle to capacity

Containers larger than 1-liter can be used but
NO MORE *than a* ***1-liter sample*** *should be obtained*

Packing and Storage

- Place bottles in cooler so ice packs and ice water can be evenly distributed around containers – blue packs or ice gel packs are preferred
- Apply the following temperature and holding time recommendations:

Fecal coliform and *Salmonella*	4 °C	24 hours
Enteric viruses	<0 °C	up to 2 weeks
Helminth	4 °C	1 month

Shipping

- Hand deliver or send by courier on ice for analysis within 6 hours of collection if possible; or
- Collect samples in the afternoon, ship on ice via express mail with overnight, morning delivery

Be SURE to read the shipping instructions brochure and documentation requirements!

Sampling Liquid Biosolids for Microbial Analysis

Attempt to select a sampling locations ***as near*** *the end of the treatment process as possible – further considering:*

Preparation

- Purge pipes to clear hardened/caked material and flush stagnant liquid prior to sampling
- If possible sterilize the spigot or pipe outlet using either:
  - a propane torch
  - 50% ethanol rinse followed by autoclaved laboratory water rinse
  - 10% diluted commercial bleach followed by multiple autoclaved laboratory rinses
- Thoroughly mix the storage tank prior to collecting a sample
- Sterilize all sampling equipment and wide-mouth autoclavable plastic or glass sample bottles prior to sample collection

Sampling

- If using a dipper, collect the sample from a tank depth ½ to ¾ of the total tank height; dippers or pre-installed hand-pump devices can be used if available
- If sampling from a pipe or spigot be sure the pipe/spigot does not come in contact with the sampling container.
- Fill a final sample container with a l-liter sample leaving ½ inch headspace
- Grab samples are recommended as part of the 503 regulations however the individual agency administering your NPDES permit should be consulted about the appropriateness of composite sampling
  - If composite samples are acceptable use sterile 150 mL or 4 oz. specimen cups for collecting subsamples for use in producing composite. Transfer 1-liter of material to container being sent to the lab for analysis.

Packing and Storing

- Place bottles in cooler so ice packs and ice water can be evenly distributed around containers – blue ice or gel ice packs are preferred
- Apply the following temperature and holding time recommendations:

Fecal coliform and *Salmonella*	4 °C	24 hours
Enteric viruses	<0 °C	up to 2 weeks
Helminth	4 °C	1 month

Shipping

- Hand deliver or send by courier on ice for analysis within 6 hours of collection if possible; or
- Collect samples in the afternoon, ship on ice via express mail with overnight, morning delivery

Be SURE to read the shipping instructions brochure and documentation requirements!

Frequency, Volume, and Number

Frequency of collecting samples is determined by the **metric tons** of biosolids produced annually at this facility.

1 × per year	Up to 290 metric tons
4 × per year	290 – 1,500 metric tons
6 × per year	1,500 – 15,000 metric tons
12 × per year	15,000 metric tons or greater

- For all three types of biosolids (liquid, cake, compost) collect **1-Liter** quantities for analysis. When collecting **liquid** leave **½" headspace** in the sample analysis container.
- Collecting **discrete (grab)** samples is recommended for microbial content analysis. When producing **"Class B, Alternative 1"** material, **seven (7)** discrete (grab) samples must be analyzed.
- For **biosolids stored longer than 30 days**, the number of samples to collect depends on the **design flow** of your facility.

Facility Design Flow	Sampling Replication
< 1 MGD	27 discrete *into* 1 composite
1 to 10 MGD	42 discrete *into* 4 composites
10 to 100 MGD	48 discrete *into* 6 composites
> 100 MGD	84 discrete *into* 12 composites

Shipping Biosolids for Microbial Analysis

The following are general guidelines for shipping procedures. BE SURE TO CHECK with your facility management concerning how biosolids in your state are defined and regulated, which can affect documentation needs.

Time Considerations

- Hand deliver or send by courier on ice for analysis **within 6 hours** of collection if possible; or
- Collect samples in the afternoon, ship on ice via overnight express service preferably with morning delivery

Packing

- Evenly distribute sample containers in cooler and use **blue ice packs**, **gel ice packs**, or double Ziploc-bagged ice/ice water to cool samples (note that your courier or shipping service may have specific restrictions that should be consulted prior to shipping)
- Be sure sample containers are tightly capped and appropriately labeled with water proof labels. Covering the label with clear tape is good for extra protection against water damage to the label.

Documentation

- Contact express service provider to obtain necessary freight shipping manifests and be sure to obtain a duplicate copy of the completed form
- Fill out chain-of-custody (COC) form for shipment, identifying all samples being shipped. Place completed forms in Ziploc bags within the pertinent shipping container. COC seals can be used on the outside of coolers/containers to provide evidence in the event of tampering during shipment
- Be sure to include a copy of the analyses requested in the cooler

IF your state regulatory agency requires a specific manifest or shipping classification for your biosolids, be sure to obtain the proper forms from your management or permitting agency.

B-8 Compliance and Public Health Bulletin Update to the Research Community

As regulatory and public health professionals it is our responsibility to ensure that measures to protect human and environmental health are implemented. As a result it is also our duty to communicate with the research community about various public health needs.

Bulletin Topic – Sampling & Handling Protocols for Microbial Analysis
This bulletin addresses research issues specific to biosolids sampling and handling practices for quality control and particularly for microbial analyses. Proper sampling and handling procedures are especially important when biosolids are to be beneficially used. Despite the presence of guidelines and protocols for sampling activities we recognize opportunities to better our understanding of ways to both improve sampling practices and to better ensure compliance occurs.

BEING IN THE KNOW

We are aware of the great diversity in processing techniques, throughputs, and technology applications among utilities and wastewater operations nationwide. As a result it may be beneficial to better understand:

1) *The relationships between treatment processes, associated biosolids quality, and requirements for rigorous sample collection and handling practices, in part to help establish effective baseline characterization of biosolids materials.*

For example, WERF Project 04-CTS-3T has demonstrated that microbial analyses are needed in addition to achieving Alternative 1 (time and temperature regime) for Class A biosolids when centrifuge dewatering is employed because of reactivation phenomenon. The application of multiple levels of quality assurance has been demonstrated to be unnecessary when using belt press dewatering as an example. Further research to establish these relationships on a process-by-process basis for industry-wide application are needed.

Benefit – Increasing our understanding of the interactions between treatment processes, biosolids product and quality, analytical variability and processed material variability is essential for optimizing sampling and handling practices. The minimum sampling requirements for one situation may be excessive for other situations. The relationship of facility-specific issues to appropriate sampling practices could improve awareness and diligence of samplers and facility managers and ultimately improve compliance and protection of environmental and human health in a cost-effective way. Research can help to determine a process for quality assurance that minimizes the frequency for which highly rigorous and repetitive practice is required without compromising compliance could help increase dedicated adherence to protocol.

2) *The need for composite versus discrete samples based on biosolids material type, the analytical method applied and end-use concerns to be addressed.*

Some currently accepted guidelines and standards require composite samples. This applies particularly for biosolids compost and other products subject to storage (i.e., > 30 days) in piles. The reason for using composite sampling protocols is to minimize the risks of missing hot spots within the bulk of biosolids material (or stored volumes) by either collecting a higher number of samples for the composite or expanding the range within stored volumes from which samples are obtained presumably resulting in more representative samples.

Results from WERF and EPA funded research projects demonstrate that analysis of multiple discrete samples as opposed to composite samples provides greater information about the variability in microbial quality of biosolids products. Pockets of biosolids with high microbial counts and thus greater potential public health risk are more easily revealed. {TEXT BOX}

Statistical analysis and research from the hazardous waste field indicates analysis of 50 or more samples are necessary to capture variability when the percentage of material that exceeds hazardous waste limits is on the order of 5%. Translating this to biosolids, in order to determine if there are pockets of material exceeding Class A or Class B criteria within a few percent of the material volume, upwards of 50 individual grab samples could be required to detect the presence of hot spots. Implementing timely microbial analyses of this number of discrete samples, within 6 hours of collection, is physically and economically impractical. Additional research is needed to determine the risk from analyzing seven samples (as described in the 503 regulations) in distribution and use of biosolids that may have contained 'hotspots' in violation of classification requirements. For Class A material, the biosolids have been subject to processes to further reduce pathogens (PFRP), therefore sampling and analysis of microbial content serves as a secondary measure of assuring material safety.

Benefit – Determining how to effectively utilize discrete sampling protocols, if applicable, would potentially help reduce sample collection and analysis costs and simplify the sampling process thus improving consistency in practice and participation in protocol among utilities and samplers.

3) *The implications of randomized versus specific locations for (liquid, cake, or compost), with respect to statistical certainty and quantitative accuracy of samples.*

EPA guidance currently specifies that biosolids sampling should occur as close in time to beneficial use as possible. However, EPA guidance also reports that sampling biosolids at a point in the treatment process where mixing occurs results in collection of the most representative samples. Therefore, in balancing practical and public health considerations, one might choose to sample biosolids at multiple points:

- Liquid – as it is being pumped into storage tanks after the final treatment process on a routine basis; periodically from the storage tank
- Cake – on a routine basis as it is leaving the dewatering device to storage; periodically from storage tanks or piles
- Compost – soon after a pile is mixed and after prescribed treatment time for windrows is reached; after the material is sent to piles from in-vessel systems; after prescribed treatment time for aerated static piles

Benefit – Research identifying specific scenarios where a randomized sampling is acceptable could be useful at facilities where greater frequency of sampling is desired and/or required. This would be due to the greater ease of sampling and potential reduction on cost of sample collection. Location-specific sample collection research can improve our understanding of how to mitigate unrepresentative biosolids quality samples and therefore be useful for enhancing compliance and ensuring end-user safety and environmental health.

4) *Evaluate what more cost-effective techniques and/or sampling tools to increase compliance rates without compromising quality assurance and minimum training levels or certifications for ensuring that sampling activities are diligently conducted by protocol.*

There is currently little up-to-date research available concerning ways to improve basic sampling tools and equipment for handling samples. For example it may be beneficial to some samplers to have a device that can be automated or calibrated to obtain exact volumes consistently. Developing and testing sampling devices to determine useful proven performance capabilities for different types of materials could contribute to higher levels of sampling consistency and thus rates of compliance (for example volumetric tools with transfer ports to lab approved storage vessels).

In addition studies to determine the effectiveness of certain forms and levels of training obtained by individuals responsible for microbial sampling and handling samples for analysis could inform ways to improve performance. The question becomes whether or not specialized training is necessary and should be required in relationship to the type of material handled, size of the operation, product quality goals, etc.

Benefit – Making available techniques and/or new equipment for sampling derived from performance-based research and specific to the type of biosolids produced and intended end-use could 1) minimize sampling errors, 2) reduce the time required for sample collection and oversight procedures, and 3) ultimately result in fewer compliance concerns.

5) Ways to improve consistency of biosolids sampling and handling practices across analytes.

Given the range of constituents for which biosolids materials are routinely analyzed, research to establish overlapping sampling protocols or those that are consistent in performance across multiple analyses would be of use. Being able to use consistent sampling practices could reduce redundancy and as a result minimize sampling costs. As such, practices that are consistent for and between nutrient, metals, pathogen, organic compounds sampling should be identified. Additional inquiries to address may include –

- Does the maturity of the biosolids material influence sampling protocol effectiveness?
- Can we define an acceptable (for all analytes) "range" of composite subsample numbers required (minimum/maximum) to mitigate error?
- Can we determine specific protocols more directly related to biosolids end-use?
- Can packing and cooling 'kits' be developed that provide more consistency in handling across facilities and material types? {TEXT BOX}

Benefit – The cost-effectiveness of biosolids sampling and time or resource requirements for collecting and handling of samples would be enhanced. As a result more consistent participation in proper practice and protocol across diverse facilities could be expected.

All aspects of improved sampling effectiveness, quality assurance measures, and compliance can benefit from on-going communications within and between utility managers, researchers, and regulatory and health professionals. {TEXT BOX}

APPENDIX C

COMMUNICATIONS ß-TEST

C-1 Communications Beta Testing Participating Facilities

Alexandria Sanitation Authority
Alexandria, VA

Bowling Green Wastewater Treatment Plant
Bowling Green, OH

City of Titusville
Titusville, FL

Orange Water and Sewer Authority
Carrboro, NC

Charleston Commissioners of Public Works
Charleston, SC

Moccasin Bend WWTP
Chattanooga Department of Public Works
Chattanooga, TN

Metropolitan Water Reclamation District of Greater Chicago
Cicero, IL

Mill Creek WWTP
Metropolitan Sewer District of Greater Cincinnati
Cincinnati, OH

Western Lake Superior Sanitary District
Duluth MN

City of Fort Worth Water Department
Fort Worth, TX

City of Fremont Water Pollution Control Center
Fremont, OH

Water Reclamation Facility
Goldsboro, NC

Western Carolina Regional Sewer Authority
Greenville, SC

City of Los Angeles Bureau of Sanitation
Los Angeles, CA

Louisville & Jefferson County Metropolitan Sewer District
Louisville, KY

Mobile Area Water and Sewer Authority
Mobile, AL

City of Raleigh Public Utilities Department
Raleigh, NC

Wastewater Treatment Plant
Port Washington, WI

Salisbury-Rowan Utilities
Salisbury, NC

City of Santa Cruz Public Works Department
Santa Cruz, CA

King County Department of Natural Resources and Parks
Seattle, WA 98104

Spartanburg Sanitary Sewer District
Spartanburg, SC

Springfield Water and Sewer Commission
Springfield, MA

Sturgeon Bay Water and Wastewater Utilities
Sturgeon Bay, WI

City of Three Rivers Wastewater Treatment Plant
Three Rivers, MI

Bayview Water Reclamation Plant
Toledo, OH

Los Angeles County Sanitation Districts
Whittier, CA

Water Corporation of Western Australia
Leederville, Western Australia

City of Ottawa Department of Public Works
Ottawa, ON Canada

C-2 Cover Letter Sent with Communications Package

WATER ENVIRONMENT RESEARCH FEDERATION
Univ. of Massachusetts/University of Wisconsin
ARCADIS U.S., Inc.

June 27, 2007

Dear Utility Representative:

As a survey and/or site visit participant in the Water Environment Research Federation project 'Biosolids Sampling and Handling Methodology for Microbial Analysis' coordinated by Dr. Sharon Long (formerly with the University of Massachusetts, now with the University of Wisconsin) your insights and feedback provide great value to this effort. In addition to the more scientific and technical elements of this work we are also focusing on the importance of communications.

The very nature of the wastewater and biosolids management industry poses unique challenges for communicating important information among and across stakeholder groups. Recognizing this, we have developed a communications strategy as part of the project and have selected a few specific audiences and communication needs from within that strategy for development of different example communications tools. A description of the various communication tools, intent and potential use of each is included herein along with various examples of products you and other interested stakeholders could utilize. Please note that within certain documents bold, underlined text represents a placeholder for information you would customize to your operation and situation.

We would appreciate your feedback about the different communications pieces created, thoughts on ways that communications could be improved upon, and other types or forms of communications you would find beneficial to your operation and industry. I will be following up with you via a telephone call to discuss this aspect of the WERF project with you during the time period July 9-11. Enclosed is also a very short set of questions for you to address either during the call, or if possible in writing and by fax to me at 919.677.9989.

Again, your assistance with this project is critical to the practical applications of the results. We appreciate a small amount of your time towards this effort.

Sincerely,

Anita R. Bahe, PhD
Project Research Consultant
919.677.9985

Cc: Dr. Sharon Long, Project Coordinator

C-3 Feedback Questionnaire

COMMUNICATIONS BETA-TEST QUESTIONNAIRE
WERF Project - Biosolids Sampling for Microbial Analysis

Goals: To assess the potential usefulness of a communications strategy and specifically certain tools or products to be made available to stakeholders.

To determine what information or issues need to be subject to improved communication and identify key audiences to be subject to targeted communication.

Content

1) Is the *general* content of the various communication product examples potentially useful?

2) What *specific* content is most important to focus communications?

3) Would the use of more visual content be beneficial in communicating with audiences you would tend to target?

Formats

1) Are the templates provided useful or appear to improve how information is conveyed?

2) How easy or complicating is it at your facility to deal with different forms of electronic files?

3) Would you prefer using simple or more visually intriguing (prettier) versions of communication pieces and products?

4) How do you currently distribute information to different groups and what do your target audiences pay attention to?

Communication Needs

1) What audiences or stakeholders need improved communication?

2) What information needs to be better communicated concerning biosolids sampling for microbial analysis and for reuse?

3) Would it be beneficial to you if your "industry" implemented a communications strategy and provided similar types of tools or products to stakeholders for your use?

C-4 Explanation of Tools Sent with Beta Test Materials

Biosolids Sampling for Microbial Analysis
Communication Needs and Opportunities

In addition to researching science-based sampling methods and technical factors influential to the effective sampling of biosolids materials for microbial analysis the research team and project sponsor (WERF) recognized the need to address various communications issues. As a result we have created an overarching strategy and framework for effective communications within which different stakeholder groups and associated communication needs are addressed. This framework is intended to be used by WERF as they see fit to encourage broad and consistent communications among and between different audiences.

Due to budget and time considerations we have developed a set of example communication pieces that can be customized and replicated. Given the research focus we opted to focus effort on communications about the purpose and practice of biosolids sampling for microbial analysis, however we also integrated issues in response to the utility surveys conducted as part of this study. As a result we have nearly completed the following communication tools.

1) Communications **from managers/operators** of wastewater facilities**:**

- *News & Notes* – this piece is a template for broad discussion about issues managers must consider in making choices and decisions about sampling practices, practical implementation of sampling protocols, health and safety, and their own information needs. This piece could be shared with a range of stakeholders internal to the operation, peers in industry, community and regulatory groups, etc
- *Internal for Staff* – this piece is designed to be very specific to staff concerning the operations commitments and requirements for quality controls and is a shortened, easy to review piece with points about sampling practices and recognition of the importance of staff accountability
- *To Community Leaders* – this piece is aimed to inform the local community leaders where the operation provides services about the practices in place for public and environmental health protection, public confidence; the concerns operators have about how to maintain quality assurances, including where other outside influences may impact activity for which community leaders should be aware.

2) Communications **specifically for samplers**:

Here there are two primary pieces including

– A *Technical Summary* within which accepted and required protocols for biosolids sampling and handling for microbial analysis are condensed into bullet points; explanations of why it is important to closely adhere to protocols; health and safety considerations specific to sampling and handling; and insights from samplers themselves into improving sampling practices onsite.

- The second item is a *Sampling Guidance Poster* which will outline and flow information in chronological fashion specific to where, when, how to obtain and then handle samples specifically for biosolids produced at a given facility. The poster will contain visuals as well as text and is intended to be strategically placed for ready reference in facilities where biosolids are sampled.

(A third fold-up quick reference pocket card may also end up being produced)

3) Communication **from regulatory or public health professionals**:

These pieces are intended to serve as an example template for specific regulatory and/or public health professionals to use in communicating needs, opportunities, and ideas with the industry, community leaders and researchers having a vested interest in biosolids sampling for protection of health and consistent quality assurances.

- *Compliance and Public Health Bulletin* – in a true bulletin style contains a synopsis of issues and concerns about ways to help make sampling practices more practical, acceptable to a wider range of facilities, science-based without being burdensome, improve consistency in practice, identifying what information exchange will help all stakeholders
- *To Researchers* – this piece, coming from compliance and health professionals to researchers evaluating ways to improve sampling practices technically, scientifically, and economically is an example of how research focus can be more clearly guided to meet multiple stakeholder needs and ensure practices in the field are in fact minimizing environmental and public health risks **(This document is not enclosed for review at this time but will be a product of this project.)**

C-5 Selected Quotes from Beta Test Questionnaire Responses

Content

Is the *general* content of the various communication product examples potentially useful?

- Concepts are useful, the contents are variable
- Content for external use too broad (not tangible) or too specific such as in News & Notes, Community Update
- Content covers all areas and operations of biosolids value chain
- Too much information/type of flyers .. only need probably 2
- Generally good especially geared toward operations personnel; not as much for the general public
- Might be useful in public relations; farmers education when needed and for contractors doing the hauling
- A few respondents felt information to deal with the public is more important than for staff; a couple commented they felt the pieces for the staff and the poster were good but for other audiences more general assumptions expressed.
- Good starting point and thought into a framework; home grown documents not widely used specific to sampling; there is room for improvement in communications within the industry
- Public officials take big picture view
- Collection frequency based on production or disposal, info presented says production where EPA says disposal/distribution

What *specific* content is most important to focus communications?

- Process changes and impacts on quality and sampling needs; better inputs from equipment vendors and system engineers
- Some in the public sector and many in the public still don't understand that biosolids are not raw sewage
- Some compliance and health officials may look this closely at the issues and practices but if not a permitting agency or someone you deal with often they don't really care that much. A few commented they felt since the 503 regs came out local officials became less involved. One individual said some in their local public offices would not know what MGD means.
- Cause, impact, concerns, cost

- The effective maintenance of regulatory compliance through the availability of consistent guidance resources from regulatory, research, and private services. The awareness of and necessity to account for variability in treatment processes. The need for training and implementation of QA/QC protocols.

Would the use of more visual content be beneficial in communicating with audiences you would tend to target?

- A little color can go along way
- Making data more useful if possible with visuals, streamline and simplify data, speak to an environmental footprint with tools
- Yes – consider power point presentations, DVD, flowcharts showing steps to meet 503
- Yes – photos, graphics, operational charts etc specific for treatment plant
- Yes – helps with comprehension
- Maybe – depends on target audience

Formats

- Some are ok, pare down further because most people don't want to read a lot
- Bullets good, better than a written memo
- If too busy or too much text can lose people's interest; bulletins and bulleted presentation can work best
- None of the formats jumped out, language good, visuals useful for both internal and external audiences
- Very informative and impressive ... show proactive attitude
- Somewhat, news and notes not really important for us but a monthly publication would be more helpful
- Maybe, depends on level of experience of target audience
- The templates are great
- Yes, the poster and News and Notes technical bulletin aimed at utility audiences are useful in communicating and reinforcing information

Respondents indicated they currently distribute information in different ways with many indicating they utilize the web (e-mail, website) as a primary tool. Some specific responses include:

- Quarterly customer newsletter, web site, Internet web site, employee newsletter … aside from biannual customer and employee surveys we do not measure response to communication pieces. Survey data indicates about 30% of our customers read our consumer newsletter

- Our website, ‘City-page’ in local newspaper, water bill insert, city letter to affected residents when necessary
- Analytical data transmitted by pdf to lone end-user who reprocesses the product along with those from other agencies
- Generally by e-mail, regular postal mail, phone, flyer … they pay attention to economic benefits to be derived and to regulatory and public health issues
- Newsletters for customers (farmers) and local elected officials, fact sheets for general public; memos and SOPs internal
- Through e-mail and verbal in meetings and with written memos; why and how it is done is what people pay attention to
- Newsletter to stakeholders; annual stakeholder meeting; annual WW report
- Internal SOP and tech memos; external post to web page
- Signs at application sites, webpage, city has section for wastewater on-line
- Public affairs use e-mail; trifold flyers by mail; sometimes stuffers in bills; solids handlers don’t have anything to hand out to customers; audiences pay attention to how safe and who is responsible party; Board asking about EDC’s and PPCP’s
- Mostly word of mouth and verbal exchange
- Send out mailings, have annual picnic with users, EMS and SOP require communications

REFERENCES

40 *Code of Federal Regulations* Part 503. 1993. Standards for the Use or Disposal of Sewage Sludge. Available through the internet: http://ecfr.gpoaccess.gov/cgi/t/text/text-idx?c=ecfr&sid=c8ce982946f3755abfbb6dd73f107ea3&tpl=/ecfrbrowse/Title40/40cfr503_main_02.tpl. Accessed September 4, 2007.

72 *Federal Register* 57. 26 March 2007. Guidelines Establishing Test Procedures for the Analysis of Pollutants; Analytical Methods for Biological Pollutants in Wastewater and Sewage Sludge; Final Rule. Pp. 14220-14233.

American Public Health Association (APHA), American Water Works Association (AWWA), and Water Environment Federation (WEF). 2005. Standard Methods for the Examination of Water and Wastewater. APHA, Washington, D.C.

American Water Works Association. 1999. Waterborne Pathogens, M48 Manual of Water Supply Practices. Awwa, Denver, CO.

ASTM. 1994. D 4994-89, Standard Practice for Recovery of Viruses from Wastewater Sludges, Annual Book of ASTM Standards, Vol. 11.02, Philadelphia, PA.

Brobst, R. 2001. Part 2 C Reference Sheets. http://www.epa.gov/region8/water/biosolids/biosolidsdown/handbook/handbook2.pdf, accessed on December 9, 2004.

CEN (Comité Européen de Normalisation). 2000. Soil Improvers and Growing Media. Sampling. EN 12579:2000.

Kenner, B.A. and H.P. Clark. 1974. Detection and Enumeration of *Salmonella* and *Pseudomonas aeruginosa*. *Journal of the Water Pollution Control Federation* **46**(9)**:** 2163-2171.

Khan, M.A. and S. Norcliff. 1982. Variability of Selected Soil Micronutrients in a Single Soil Series in Berkshire, England. *Journal of Soil Science* **33**(4): 763-770.

Lambkin, D., S. Nortcliff, and T. White. 2004. The Importance of Precision in Sampling Sludges, Biowastes and Treated Soils in a Regulatory Framework. *Trends in Analytical Chemistry* **23**(10-11): 704-715.

Lamé, F., T. Honders, D. Derkson, and M. Gadella. 2005. Validated Sampling Strategy for Assessing Contaminants in Soil Stockpiles, *Environmental Pollution*, **134**(1): 5-11.

Laraia, R., L. Cortellini, and F. DePoli. 2002. Sampling Protocols for Ensuring Legal Compliance and Standardisation: The case of Italy. Biological treatment of biodegradable waste - Technical aspects. Proceedings of the JRC Workshop in Brussels, April 8-10 (2002), European Commission, Brussels.Meckes, M. 2007. Personal communications, U.S. Environmental Protection Agency, Cincinnati, OH.

Mathar, R. and B. Rauhut. 1983. Sampling Procedures for Bulk Solids. *Bulk Solids Handling*, **3**(1): 189-190.

NEIWPCC. 2006. The Wastewater Treatment Plant Operators Guide to Biosolids Sampling Plans. Available through NEIWPCC website: http://www.neiwpcc.org/biosamplingguide.asp.

New Jersey Administrative Code. NJAC 7:18. Available through the internet: http://www.state.nj.us/dep/oqa/docs/final%20Copy%20of%20REGSwithair.pdf.

New Jersey Department of Environmental Protection. 2005. NJDEP Field Sampling Procedures Manual. Available through the internet: http://www.state.nj.us/dep/srp/guidance/fspm/pdf/fsmp2005.pdf.

Parkhurst, D.F. 1984. Optimal Sampling Geometry for Hazardous Waste Sites. *Environmental Science and Technology* **18**(7): 521-523

Reiners, E. and P. Mohrhauer. 1994. Devices and Equipment for Sampling and Sample Preparation of Bulk Materials. *Aufbereitungs-Technik* **35**(9): 457-467.

Tinklenberg, G. and L. Smith. 1990. The Criticality of Sampling and Quality Control for Hazardous Waste Testing. *Journal of Protective Coatings & Linings* **7**(4): 36-44.

U.S. EPA. 1989. POTW Sludge Sampling and Analysis Guidance Document. EPA 833-B-89-100, Office of Water, Washington, D.C. Available through the internet: http://www.epa.gov/npdes/pubs/owm012.pdf.

U.S. EPA. 1993. Preparing Sewage Sludge for Land Application or Surface Disposal. EPA831B/93/002a, Office of Water, Washington, D.C. Available through the National Service Center for Environmental Publications (NSCEP): http://www.epa.gov/nscep/.

U.S. EPA. 1993a. Sewage Sludge Sampling Techniques. Guidance video, March 1993.

U.S. EPA. 1994. Plain English Guide to the EPA Part 503 Biosolids Rule. EPA 832/R-93/003, Office of Wastewater Management, Washington, D.C.

U.S. EPA. 1995. A Guide to the Biosolids Risk Assessments for the EPA Part 503 Rule, EPA-832-B-93-005, EPA Office of Wastewater Management, Washington, D.C.

U.S. EPA. 1999. Control of Pathogens and Vector Attraction in Sewage Sludge, EPA-625-R-92-013, EPA Office of Research and Development: Washington, D.C.

U.S. EPA. 1999a. Biosolids Generation, Use, and Disposal in the United States, EPA-530 R-99-009, Office of Solid Waste and Emergency Response: Washington, D.C. Available at: http://www.epa.gov/epaoswer/non-hw/compost/biosolid.pdf

U.S. EPA. 2000. Guide to Field Storage of Biosolids and Other Organic By-Products Used in Agriculture and for Soil Resource Management, EPA-832-B-00-007, Office of Wastewater Management, Washington, D.C.

U.S. EPA. 2000a. Biosolids Technology Fact Sheet: Land Application of Biosolids, EPA 832-F-00-064, EPA Office of Water, Washington, D.C. Available at: http://www.epa.gov/OWM/mtb/land_application.pdf

U.S. EPA. 2001. Method 1602: Male-Specific (F^+) and Somatic Coliphage in Water by Single Agar Layer (SAL) Procedure; EPA-821-R-01-029; EPA Office of Water, Washington, D.C.

U.S. EPA. 2003. White House Document (Revised July 2003), EPA 625/R-92/013, Office of Research and Development, Washington, D.C. Available through the EPA website: http://www.epa.gov/ord/ NRMRL/Pubs/1992/625R92013.pdf.

U.S, EPA. 2004. SW-846, Test Methods for Evaluating Solid Waste, Physical/Chemical Methods, Volume 2, Chapter 9 – Sampling Plan, version 6, Nov. 2004. http://www.epa.gov/epaoswer/ hazwaste/test/pdfs/chap9.pdf (accessed March 13, 2005).

U.S. EPA. 2006. Method 1680: Fecal Coliforms in Sewage Sludge (Biosolids) by Multiple-Tube Fermentation using Lauryl Tryptose Broth (LTB) and EC Medium, EPA-821-R-04-026; EPA Office of Water, Washington, D.C.

U.S. EPA. 2006a. Method 1681: Fecal Coliforms in Sewage Sludge (Biosolids) by Multiple-Tube Fermentation using A-1 Medium, EPA-821-R-06-013, EPA Office Water, Washington, D.C.

U.S. EPA. 2006b. Method 1682: *Salmonella* in Sewage Sludge (Biosolids) by Modified Semisolid Rappaport-Vassiliadis (MSRV) Medium, EPA-821-R-06-014, EPA Office of Water, Washington, D.C.

U.S. EPA. 2006c. Assessment of the Effects of Holding Time on Fecal Coliform and *Salmonella* Concentrations in Biosolids, EPA-821-R-07-003, EPA Office of Water, Washington, D.C.

U.S. FDA (Food and Drug Administration). 1992. Bad Bug Book, Chapter 1 *Salmonella* spp. http://www.cfsan.fda.gov/~mow/chap1.html, accessed on October 10, 2007.

van Elsas, J.D. and K. Smalla. 1997. Methods for Sampling Soil Microbes. *Manual of Environmental Microbiology*, C.J. Hurst *et al.* (eds). ASM Press, Washington, D.C. pp. 383-390.

Water Environment Federation. 1996. WEF MOP OM-1. Wastewater Sampling for Process and Quality Control. Available for purchase through: http://www.e-wef.org/timssnet/static/OM/ MF2010.htm.

Yanko, W.A. 1987. Occurrence of Pathogens in Distribution and Marketing Municipal Sludges. Report No. EPA/600/1-87/014. National Technical Information Service, Springfield, VA.

Yates, M. 2005. Viruses in Water: Sources and Monitoring. In Proceedings of the 2005 AWWA Source Water Protection Symposium, January 23-26 (2005), Palm Beach Gardens, FL.

WERF SUBSCRIBERS

WASTEWATER UTILITY

Alabama
Montgomery Water Works & Sanitary Sewer Board

Alaska
Anchorage Water & Wastewater Utility

Arizona
Avondale, City of
Glendale, City of, Utilities Department
Mesa, City of
Peoria, City of
Phoenix Water Services Dept.
Pima County Wastewater Management
Safford, City of
Tempe, City of

Arkansas
Little Rock Wastewater Utility

California
Central Contra Costa Sanitary District
Corona, City of
Crestline Sanitation District
Delta Diablo Sanitation District
Dublin San Ramon Services District
East Bay Dischargers Authority
East Bay Municipal Utility District
El Dorado Irrigation District
Fairfield-Suisun Sewer District
Fresno Department of Public Utilities
Inland Empire Utilities Agency
Irvine Ranch Water District
Las Gallinas Valley Sanitary District
Las Virgenes Municipal Water District
Livermore, City of
Los Angeles, City of
Los Angeles County, Sanitation Districts of
Napa Sanitation District
Novato Sanitary District
Orange County Sanitation District
Palo Alto, City of
Riverside, City of
Sacramento Regional County Sanitation District
San Diego Metropolitan Wastewater Department, City of
San Francisco, City & County of
San Jose, City of
Santa Barbara, City of
Santa Cruz, City of
Santa Rosa, City of
South Bayside System Authority
South Coast Water District
South Orange County Wastewater Authority
Stege Sanitary District
Sunnyvale, City of
Union Sanitary District
West Valley Sanitation District

Colorado
Aurora, City of
Boulder, City of
Greeley, City of
Littleton/Englewood Water Pollution Control Plant
Metro Wastewater Reclamation District, Denver

Connecticut
Greater New Haven WPCA
Stamford, City of

District of Columbia
District of Columbia Water & Sewer Authority

Florida
Broward, County of
Fort Lauderdale, City of
Jacksonville Electric Authority (JEA)
Miami-Dade Water & Sewer Authority
Orange County Utilities Department
Pinellas, County of
Reedy Creek Improvement District
Seminole County Environmental Services
St. Petersburg, City of
Tallahassee, City of
Tampa, City of
Toho Water Authority
West Palm Beach, City of

Georgia
Atlanta Department of Watershed Management
Augusta, City of
Clayton County Water Authority
Cobb County Water System
Columbus Water Works
Fulton County
Gwinnett County Department of Public Utilities
Savannah, City of

Hawaii
Honolulu, City & County of

Idaho
Boise, City of

Illinois
Decatur, Sanitary District of
Greater Peoria Sanitary District
Kankakee River Metropolitan Agency
Metropolitan Water Reclamation District of Greater Chicago
Wheaton Sanitary District

Iowa
Ames, City of
Cedar Rapids Wastewater Facility
Des Moines, City of
Iowa City

Kansas
Johnson County Wastewater
Lenexa, City of
Unified Government of Wyandotte County/ Kansas City, City of

Kentucky
Louisville & Jefferson County Metropolitan Sewer District

Louisiana
Sewerage & Water Board of New Orleans

Maine
Bangor, City of
Portland Water District

Maryland
Anne Arundel County Bureau of Utility Operations
Howard County Bureau of Utilities
Washington Suburban Sanitary Commission

Massachusetts
Boston Water & Sewer Commission
Massachusetts Water Resources Authority (MWRA)
Upper Blackstone Water Pollution Abatement District

Michigan
Ann Arbor, City of
Detroit, City of
Holland Board of Public Works
Saginaw, City of
Wayne County Department of Environment
Wyoming, City of

Minnesota
Rochester, City of
Western Lake Superior Sanitary District

Missouri
Independence, City of
Kansas City Missouri Water Services Department
Little Blue Valley Sewer District
Metropolitan St. Louis Sewer District

Nebraska
Lincoln Public Works and Utilities Department

Nevada
Henderson, City of
Las Vegas, City of
Reno, City of

New Jersey
Bergen County Utilities Authority
Ocean County Utilities Authority
Passaic Valley Sewerage Commissioners

New York
New York City Department of Environmental Protection

North Carolina
Charlotte/Mecklenburg Utilities
Durham, City of
Metropolitan Sewerage District of Buncombe County
Orange Water & Sewer Authority
University of North Carolina, Chapel Hill

Ohio
Akron, City of
Butler County Department of Environmental Services
Columbus, City of
Metropolitan Sewer District of Greater Cincinnati
Montgomery, County of
Northeast Ohio Regional Sewer District
Summit, County of

Oklahoma
Oklahoma City Water & Wastewater Utility Department
Tulsa, City of

Oregon
Albany, City of
Clean Water Services
Eugene, City of
Gresham, City of
Portland, City of Bureau of Environmental Services
Water Environment Services

Pennsylvania
Hemlock Municipal Sewer Cooperative (HMSC)
Philadelphia, City of
University Area Joint Authority

South Carolina
Charleston Water System
Mount Pleasant Waterworks & Sewer Commission
Spartanburg Water

Tennessee
Cleveland Utilities
Murfreesboro Water & Sewer Department
Nashville Metro Water Services

Texas
Austin, City of
Dallas Water Utilities
Denton, City of
El Paso Water Utilities

Fort Worth, City of
Houston, City of
San Antonio Water System
Trinity River Authority

Utah

Salt Lake City Corporation

Virginia

Alexandria Sanitation Authority
Arlington, County of
Fairfax County
Hampton Roads Sanitation District
Hanover, County of
Henrico, County of
Hopewell Regional Wastewater Treatment Facility
Loudoun Water
Prince William County Service Authority
Richmond, City of
Rivanna Water & Sewer Authority

Washington

Everett, City of
King County Department of Natural Resources
Seattle Public Utilities
Sunnyside, Port of
Yakima, City of

Wisconsin

Green Bay Metro Sewerage District
Kenosha Water Utility
Madison Metropolitan Sewerage District
Milwaukee Metropolitan Sewerage District
Racine, City of
Sheboygan Regional Wastewater Treatment
Wausau Water Works

Australia

ACTEW (Ecowise)
South Australian Water Corporation
South East Water Limited
Sydney Water Corporation
Water Corporation of Western Australia

Canada

Edmonton, City of/Edmonton Waste Management Centre of Excellence
Lethbridge, City of
Regina, City of, Saskatchewan
Toronto, City of, Ontario
Winnipeg, City of, Manitoba

New Zealand

Watercare Services Limited

United Kingdom

Anglian Water Services, Ltd.

STORMWATER UTILITY

California

Fresno Metropolitan Flood Control District
Los Angeles, City of, Department of Public Works
Monterey, City of
San Francisco, City & County of
Santa Rosa, City of
Sunnyvale, City of

Colorado

Aurora, City of
Boulder, City of

Florida

Orlando, City of

Iowa

Cedar Rapids Wastewater Facility
Des Moines, City of

Kansas

Overland Park, City of

Kentucky

Louisville & Jefferson County Metropolitan Sewer District

Maine

Portland Water District

North Carolina

Charlotte, City of, Stormwater Services

Pennsylvania

Philadelphia, City of

Tennessee

Chattanooga Stormwater Management

Texas

Harris County Flood Control District, Texas

Washington

Bellevue Utilities Department
Seattle Public Utilities

STATE

Connecticut Department of Environmental Protection
Kansas Department of Health & Environment
New England Interstate Water Pollution Control Commission (NEIWPCC)
Ohio River Valley Sanitation Commission
Urban Drainage & Flood Control District, CO

CORPORATE

ADS Environmental Services
Advanced Data Mining International
AECOM
Alan Plummer & Associates
Alpine Technology Inc.
Aqua-Aerobic Systems Inc.
Aquateam–Norwegian Water Technology Centre A/S
ARCADIS
Associated Engineering
Bernardin Lochmueller & Associates
Black & Veatch
Blue Water Technologies, Inc.
Brown & Caldwell
Burgess & Niple, Ltd.
Burns & McDonnell
CABE Associates Inc.
The Cadmus Group
Camp Dresser & McKee Inc.
Carollo Engineers Inc.
Carpenter Environmental Associates Inc.
CET Engineering Services
CH2M HILL
CRA Infrastructure & Engineering
CONTECH Stormwater Solutions
D&B/Guarino Engineers, LLC
Damon S. Williams Associates, LLC
Ecovation
EMA Inc.
Environmental Operating Solutions, Inc.
Environ/The ADVENT Group, Inc.
Fay, Spofford, & Thorndike Inc.
Freese & Nichols, Inc.
ftn Associates Inc.
Gannett Fleming Inc.
Garden & Associates, Ltd.
Geosyntec Consultants
GHD
Greeley and Hansen LLC
Hazen & Sawyer, P.C.
HDR Engineering Inc.
HNTB Corporation
Hydromantis Inc.
HydroQual Inc.
Infilco Degremont Inc.
Jacques Whitford NAWE, Inc.
Jason Consultants LLC Inc.
Jordan, Jones, & Goulding Inc.
KCI Technologies Inc.
Kelly & Weaver, P.C.
Kennedy/Jenks Consultants
KMK Consultants
Larry Walker Associates
Limno-Tech Inc.
The Low Impact Development Center Inc.
Malcolm Pirnie Inc.
Material Matters
McKim & Creed
MPR Engineering Corporation, Inc.
MWH
NTL Alaska, Inc.
O'Brien & Gere Engineers Inc.
Odor & Corrosion Technology Consultants Inc.
Paradigm Environmental Technologies, Inc.
Parametrix Inc.
Parsons
Post, Buckley, Schuh & Jernigan
Praxair, Inc.
Ring Industrial Group
RMC Water & Environment
Ross & Associates Ltd.
SAIC
Siemens Water Technologies
The Soap & Detergent Association
Smith & Loveless, Inc.
Southeast Environmental Engineering, LLC
Stearns & Wheler, LLC
Stone Environmental Inc.
Stratus Consulting Inc.
Synagro Technologies Inc.
Tetra Tech Inc.
Trojan Technologies Inc.
Trussell Technologies, Inc.
Uni-Bell PVC Pipe Association
URS Corporation
Westin Engineering Inc.
Wright Water Engineers
Zoeller Pump Company

INDUSTRY

American Electric Power
American Water
Anglian Water Services, Ltd.
Chevron Energy Technology
The Coca-Cola Company
Dow Chemical Company
DuPont Company
Eastman Chemical Company
Eli Lilly & Company
Johnson & Johnson
Merck & Company Inc.
Procter & Gamble Company
Suez Environnment
United Utilities North West (UUNW)
United Water Services LLC

Note: List as of 2/21/09

IWA Publishing's authorised EU representative for General Product Safety Regulations is Diane D'Arras, 15 rue Duret, 75116 Paris, France, e-mail: safety@iwap.co.uk.

Printed and bound by CPI Group (UK) Ltd, Croydon, CR0 4YY
12/05/2026
02108876-0017